SPHERICAL TRIGONOMETRY

For the Use of Colleges and Schools
With Numerous Examples

I. TODHUNTER, M.A., F.R.S.,
HONORARY FELLOW OF ST JOHN'S COLLEGE,
CAMBRIDGE.

Published by

Hawk Press
4836/24, Ansari Road, Daryaganj
New Delhi - 110 002
Phones : 9643330713, 011-23278618, 011-35676207
E-mail : thehawkpress@gmail.com
www.thehawkpress.com

ISBN : 978-93-93971-59-3

PREFACE

The present work is constructed on the same plan as my treatise on Plane Trigonometry, to which it is intended as a sequel; it contains all the propositions usually included under the head of Spherical Trigonometry, together with a large collection of examples for exercise. In the course of the work reference is made to preceding writers from whom assistance has been obtained; besides these writers I have consulted the treatises on Trigonometry by Lardner, Lefebure de Fourcy, and Snowball, and the treatise on Geometry published in the Library of Useful Knowledge. The examples have been chiefly selected from the University and College Examination Papers.

In the account of Napier's Rules of Circular Parts an explanation has been given of a method of proof devised by Napier, which seems to have been overlooked by most modern writers on the subject. I have had the advantage of access to an unprinted Memoir on this point by the late R. L. Ellis of Trinity College; Mr Ellis had in fact rediscovered for himself Napier's own method. For the use of this Memoir and for some valuable references on the subject I am indebted to the Dean of Ely.

Considerable labour has been bestowed on the text in order to render it comprehensive and accurate, and the examples have all been carefully verified; and thus I venture to hope that the work will be found useful by Students and Teachers.

I. TODHUNTER.

ST JOHN'S COLLEGE,
August 15, 1859.

In the third edition I have made some additions which I hope will be found valuable. I have considerably enlarged the discussion on the connexion of Formulæ in Plane and Spherical Trigonometry; so as to include an account of the properties in Spherical Trigonometry which are analogous to those of the Nine Points Circle in Plane Geometry. The mode of investigation is more elementary than those hitherto employed; and perhaps some of the results are new. The fourteenth Chapter is almost entirely original, and may deserve attention from the nature of the propositions themselves and of the demonstrations which are given.

CAMBRIDGE,
July, 1871.

CONTENTS.

I

GREAT AND SMALL CIRCLES.

1. A SPHERE is a solid bounded by a surface every point of which is equally distant from a fixed point which is called the *centre* of the sphere. The straight line which joins any point of the surface with the centre is called a *radius*. A straight line drawn through the centre and terminated both ways by the surface is called a *diameter*.

2. *The section of the surface of a sphere made by any plane is a circle.*

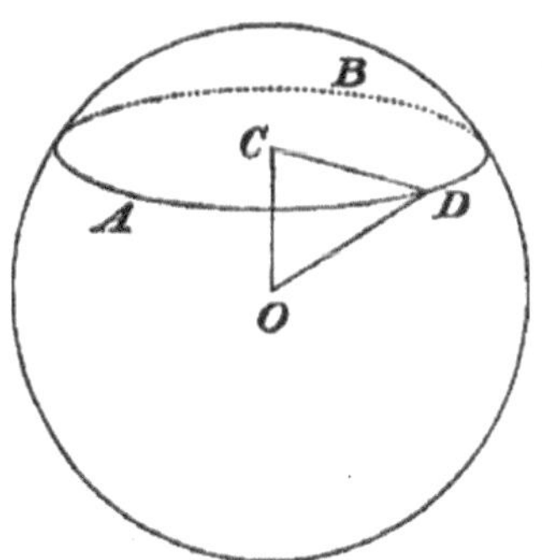

Let AB be the section of the surface of a sphere made by any plane, O the centre of the sphere. Draw OC perpendicular to the plane; take any point D in the section and join OD, CD. Since OC is perpendicular to the plane, the angle OCD is a right angle; therefore $CD = \sqrt{(OD^2 - OC^2)}$. Now O and C are fixed points, so that OC is constant; and OD is constant, being the radius of the sphere; hence CD is constant. Thus all points in the plane section are equally distant from the fixed point C; therefore the section is a circle of which C is the centre.

3. The section of the surface of a sphere by a plane is called a *great circle* if the plane passes through the centre of the sphere, and a *small circle* if the plane does not pass through the centre of the sphere. Thus the radius of a great circle is equal to the radius of the sphere.

4. Through the centre of a sphere and any two points on the surface a plane can be drawn; and only one plane can be drawn, except when the two points are the extremities of a diameter of the sphere, and then an infinite number of such planes can be drawn. Hence only one great circle can be drawn through two given points on the surface of a sphere, except when the points are the extremities of a diameter of the sphere. When only one great circle can be drawn through two given points, the great circle is unequally divided at the two points; we shall for brevity speak of the shorter of the two arcs as *the* arc of a great circle joining the two points.

5. The *axis* of any circle of a sphere is that diameter of the sphere which is perpendicular to the plane of the circle; the extremities of the axis are called the *poles* of the circle. The poles of a great circle are equally distant from the plane of the circle. The poles of a small circle are not equally distant from the plane of the circle; they may be called respectively the *nearer* and *further* pole; sometimes the nearer pole is for brevity called *the* pole.

6. *A pole of a circle is equally distant from every point of the circumference of the circle.*

Let O be the centre of the sphere, AB any circle of the sphere, C the centre of the circle, P and P' the poles of the circle. Take any point D in the circumference of the circle; join CD, OD, PD. Then $PD = \sqrt{(PC^2 + CD^2)}$; and PC and CD are constant, therefore PD is constant. Suppose a great circle to pass through the points P and D; then the chord PD is constant, and therefore the arc of a great circle intercepted between P and D is constant for all positions of D on the circle AB.

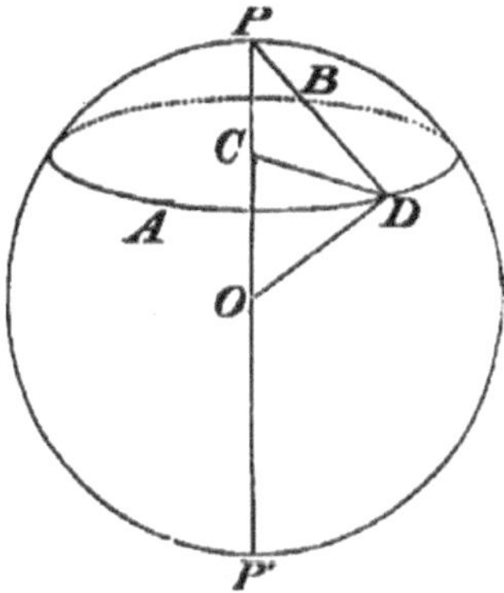

Thus the distance of a pole of a circle from every point of the circumference of the circle is constant, whether that distance be measured by the straight line joining the points, or by the arc of a great circle intercepted between the points.

7. *The arc of a great circle which is drawn from a pole of a great circle to any point in its circumference is a quadrant.*

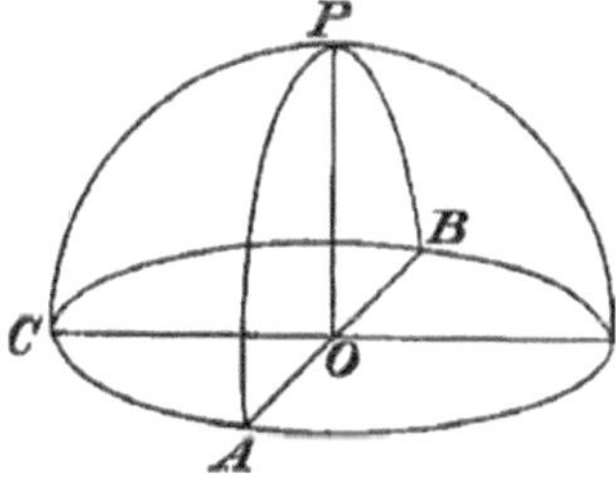

Let P be a pole of the great circle ABC; then the arc PA is a quadrant.

For let O be the centre of the sphere, and draw PO. Then PO is at right angles to the plane ABC, because P is the pole of ABC, therefore POA is a right angle, and the arc PA is a quadrant.

8. *The angle subtended at the centre of a sphere by the arc of a great circle which joins the poles of two great circles is equal to the inclination of the planes of the great circles.*

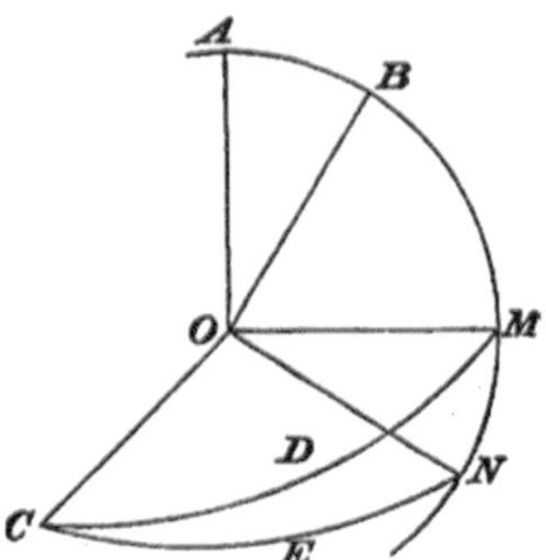

Let O be the centre of the sphere, CD, CE the great circles intersecting at C, A and B the poles of CD and CE respectively.

Draw a great circle through A and B, meeting CD and CE at M and N respectively. Then AO is perpendicular to OC, which is a straight line in the plane OCD; and BO is perpendicular to OC, which is a straight line in the plane OCE; therefore OC is perpendicular to the plane AOB (Euclid, XI. 4); and therefore OC is perpendicular to the straight lines OM and ON, which are in the plane AOB. Hence MON is the angle of inclination of the planes OCD and OCE. And the angle

$$AOB = AOM - BOM = BON - BOM = MON.$$

9. By the angle between two great circles is meant *the angle of inclination of the planes of the circles.* Thus, in the figure of the preceding Article, the angle between the great circles CD and CE is the angle MON.

In the figure to Art. 6, since PO is perpendicular to the plane ACB, every plane which contains PO is at right angles to the plane ACB. Hence the angle between the plane of any circle and the plane of a great circle which passes through its poles is a right angle.

10. *Two great circles bisect each other.*

For since the plane of each great circle passes through the centre of the sphere, the line of intersection of these planes is a diameter of the sphere, and therefore also a diameter of each great circle; therefore the great circles are bisected at the points where they meet.

11. *If the arcs of great circles joining a point* P *on the surface of a sphere with two other points* A *and* C *on the surface of the sphere, which are not at opposite extremities of a diameter, be each of them equal to a quadrant,* P *is a pole of the great circle through* A *and* C. (See the figure of Art. 7.)

For suppose PA and PC to be quadrants, and O the centre of the sphere; then since PA and PC are quadrants, the angles POC and POA are right angles. Hence PO is at right angles to the plane AOC, and P is a pole of the great circle AC.

12. Great circles which pass through the poles of a great circle are called *secondaries* to that circle. Thus, in the figure of Art. 8 the point C is a pole of $ABMN$, and therefore CM and CN are parts of secondaries to $ABMN$. And the angle between CM and CN is measured by MN; that is, *the angle between any two great circles is measured by the arc they intercept on the great circle to which they are secondaries.*

13. *If from a point on the surface of a sphere there can be drawn two arcs of great circles, not parts of the same great circle, the planes of which are at right angles to the plane of a given circle, that point is a pole of the given circle.*

For, since the planes of these arcs are at right angles to the plane of the given circle, the line in which they intersect is perpendicular to the plane of the given circle, and is therefore the axis of the given circle; hence the point from which the arcs are drawn is a pole of the circle.

14. *To compare the arc of a small circle subtending any angle at the centre of the circle with the arc of a great circle subtending the same angle at its centre.*

Let ab be the arc of a small circle, C the centre of the circle, P the pole of the circle, O the centre of the sphere. Through P draw the great circles PaA and PbB, meeting the great circle of which P is a pole, at A and B respectively; draw Ca, Cb, OA, OB. Then Ca, Cb, OA, OB are all perpendicular to OP, because the planes aCb and AOB are perpendicular to OP; therefore Ca is parallel to OA, and Cb is parallel to OB. Therefore the angle aCb = the angle AOB (Euclid, XI. 10). Hence,

$$\frac{\text{arc}\, ab}{\text{radius}\, Ca} = \frac{\text{arc}\, AB}{\text{radius}\, OA}, \quad (\textit{Plane Trigonometry}, \text{Art. } 18);$$

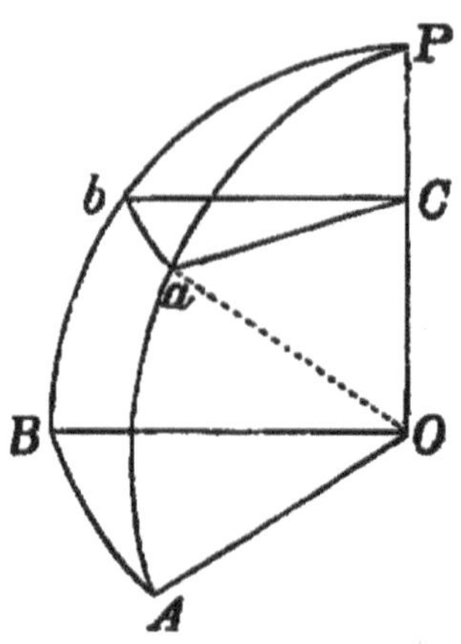

therefore, $$\frac{\text{arc}\, ab}{\text{arc}\, AB} = \frac{Ca}{OA} = \frac{Ca}{Oa} = \sin POa.$$

II

SPHERICAL TRIANGLES.

15. Spherical Trigonometry investigates the relations which subsist between the angles of the plane faces which form a solid angle and the angles at which the plane faces are inclined to each other.

16. Suppose that the angular point of a solid angle is made the centre of a sphere; then the planes which form the solid angle will cut the sphere in arcs of great circles. Thus a figure will be formed on the surface of the sphere which is called a *spherical triangle* if it is bounded by *three* arcs of great circles; this will be the case when the solid angle is formed by the meeting of *three* plane angles. If the solid angle be formed by the meeting of *more than three* plane angles, the corresponding figure on the surface of the sphere is bounded by more than three arcs of great circles, and is called a *spherical polygon.*

17. The three arcs of great circles which form a spherical triangle are called the *sides* of the spherical triangle; the angles formed by the arcs at the points where they meet are called the *angles* of the spherical triangle. (See Art. 9.)

18. Thus, let O be the centre of a sphere, and suppose a solid angle formed at O by the meeting of three plane angles. Let AB, BC, CA be the arcs of great circles in which the planes cut the sphere; then ABC is a spherical triangle, and

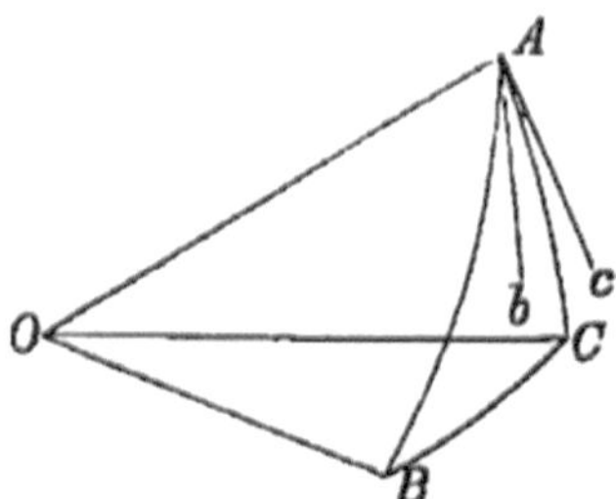

the arcs AB, BC, CA are its sides. Suppose Ab the tangent at A to the arc AB, and Ac the tangent at A to the arc AC, the tangents being drawn from A *towards* B and C respectively; then the angle bAc is one of the angles of the spherical triangle. Similarly angles formed in like manner at B and C are the other angles of the spherical triangle.

19. The principal part of a treatise on Spherical Trigonometry consists of theorems relating to spherical triangles; it is therefore necessary to obtain an accurate conception of a spherical triangle and its parts.

It will be seen that what are called *sides* of a spherical triangle are really *arcs* of great circles, and these arcs are proportional to the three plane angles which form the solid angle corresponding to the spherical triangle. Thus, in the figure of the preceding Article, the arc AB forms one side of the spherical triangle ABC, and the plane angle AOB is measured by the fraction $\frac{\text{arc}\,AB}{\text{radius}\,OA}$; and thus the arc AB is proportional to the angle AOB so long as we keep to the same sphere.

The *angles* of a spherical triangle are the inclinations of the plane faces which form the solid angle; for since Ab and Ac are both perpendicular to OA, the angle bAc is the angle of inclination of the planes OAB and OAC.

20. The letters A, B, C are generally used to denote the *angles* of a spherical triangle, and the letters a, b, c are used to denote the *sides*. As in the case of plane triangles, A, B, and C may be used to denote the numerical values of the angles expressed in *terms of any unit*, provided we understand distinctly what the unit is. Thus, if the angle C be a right angle, we may say that $C = 90°$, or that $C = \frac{\pi}{2}$, according as we adopt for the unit a degree or the angle subtended at the centre by an arc equal to the radius. So also, as the sides of a spherical triangle are proportional to the angles subtended at the centre of the sphere, we may use a, b, c to denote the numerical values of those angles in terms of any unit. We shall usually suppose both the angles and sides of a spherical triangle expressed in *circular measure.* (*Plane Trigonometry*, Art. 20.)

21. In future, unless the contrary be distinctly stated, any arc drawn on the surface of a sphere will be supposed to be an arc of a *great* circle.

22. In spherical triangles each side is restricted to be less than a semicircle; this is of course a *convention*, and it is adopted because it is found convenient.

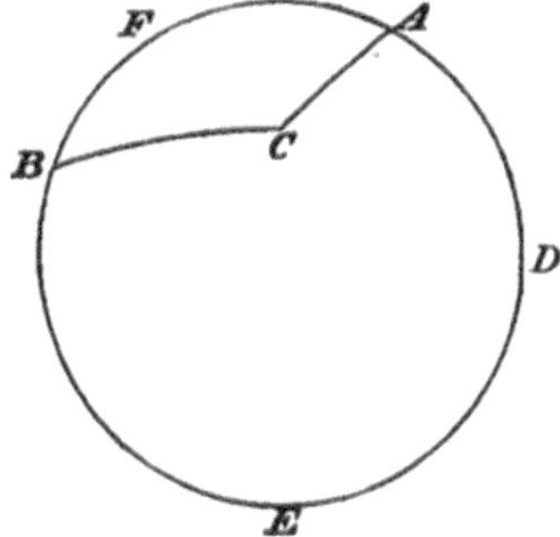

Thus, in the figure, the arc $ADEB$ is greater than a semicircumference, and we might, if we pleased, consider $ADEB$, AC, and BC as forming a triangle, having its angular points at A, B, and C. But we agree to exclude such triangles from our consideration; and the triangle having its angular points at A, B, and C, will be understood to be that formed by AFB, BC, and CA.

23. From the restriction of the preceding Article it will follow that *any angle of a spherical triangle is less than two right angles.*

For suppose a triangle formed by BC, CA, and $BEDA$, having the angle BCA greater than two right angles. Then suppose D to denote the point at which the arc BC, if produced, will meet AE; then BED is a semicircle by Art. 10, and therefore BEA is greater than a semicircle; thus the proposed triangle is not one of those which we consider.

III

SPHERICAL GEOMETRY.

24. The relations between the sides and angles of a Spherical Triangle, which are investigated in treatises on Spherical Trigonometry, are chiefly such as involve the *Trigonometrical Functions* of the sides and angles. Before proceeding to these, however, we shall collect, under the head of Spherical Geometry, some theorems which involve the sides and angles *themselves*, and not their trigonometrical ratios.

25. *Polar triangle.* Let ABC be any spherical triangle, and let the points A', B', C' be those poles of the arcs BC, CA, AB respectively which lie on the

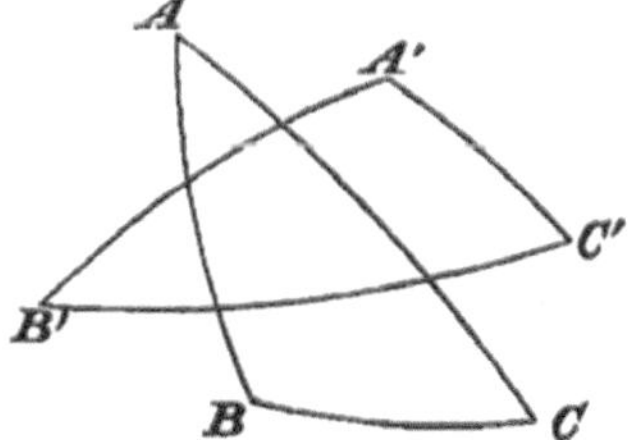

same sides of them as the opposite angles A, B, C; then the triangle $A'B'C'$ is said to be the *polar triangle* of the triangle ABC.

Since there are two poles for each side of a spherical triangle, *eight* triangles can be formed having for their angular points poles of the sides of the given triangle; but there is only one triangle in which these poles A', B', C' lie towards the same parts with the corresponding angles A, B, C; and this is the triangle which is known under the name of the *polar triangle.*

The triangle ABC is called the *primitive* triangle with respect to the triangle $A'B'C'$.

26. *If one triangle be the polar triangle of another, the latter will be the polar triangle of the former.*

Let ABC be any triangle, $A'B'C'$ the polar triangle: then ABC will be the polar triangle of $A'B'C'$.

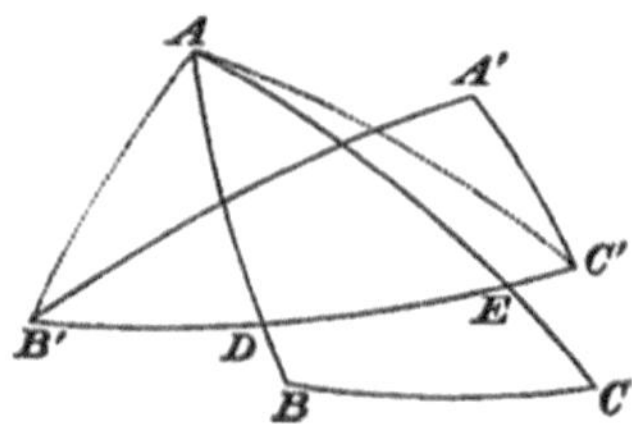

For since B' is a pole of AC, the arc AB' is a quadrant, and since C' is a pole of BA, the arc AC' is a quadrant (Art. 7); therefore A is a pole of $B'C'$ (Art. 11). Also A and A' are on the same side of $B'C'$; for A and A' are by hypothesis on the same side of BC, therefore $A'A$ is less than a quadrant; and since A is a pole of $B'C'$, and AA' is less than a quadrant, A and A' are on the same side of $B'C'$.

Similarly it may be shewn that B is a pole of $C'A'$, and that B and B' are on the same side of $C'A'$; also that C is a pole of $A'B'$, and that C and C' are on the same side of $A'B'$. Thus ABC is the polar triangle of $A'B'C'$.

27. *The sides and angles of the polar triangle are respectively the supplements of the angles and sides of the primitive triangle.*

For let the arc $B'C'$, produced if necessary, meet the arcs AB, AC, produced if necessary, at the points D and E respectively; then since A is a pole of $B'C'$, the spherical angle A is measured by the arc DE (Art. 12). But $B'E$ and $C'D$ are each quadrants; therefore DE and $B'C'$ are together equal to a semicircle; that is, the angle subtended by $B'C'$ at the centre of the sphere is the supplement of the angle A. This we may express for shortness thus; $B'C'$ is the supplement of A. Similarly it may be shewn that $C'A'$ is the supplement of B, and $A'B'$ the supplement of C.

And since ABC is the polar triangle of $A'B'C'$, it follows that BC, CA, AB are respectively the supplements of A', B', C'; that is, A', B', C' are respectively the supplements of BC, CA, AB.

From these properties a primitive triangle and its polar triangle are sometimes called *supplemental triangles.*

Thus, if A, B, C, a, b, c denote respectively the angles and the sides of a spherical triangle, all expressed in circular measure, and A', B', C', a', b', c' those of the polar triangle, we have

$$A' = \pi - a, \qquad B' = \pi - b, \qquad C' = \pi - c,$$
$$a' = \pi - A, \qquad b' = \pi - B, \qquad c' = \pi - C.$$

28. The preceding result is of great importance; for if any general theorem be demonstrated with respect to the sides and the angles of any spherical triangle it holds of course for the polar triangle also. *Thus any such theorem will remain true when the angles are changed into the supplements of the corresponding sides and the sides into the supplements of the corresponding angles.* We shall see several examples of this principle in the next Chapter.

29. *Any two sides of a spherical triangle are together greater than the third side.* (See the figure of Art. 18.)

For any two of the three plane angles which form the solid angle at O are together greater than the third (Euclid, XI. 20). Therefore any two of the arcs AB, BC, CA, are together greater than the third.

From this proposition it is obvious that any side of a spherical triangle is greater than the difference of the other two.

30. *The sum of the three sides of a spherical triangle is less than the circumference of a great circle.* (See the figure of Art. 18.)

For the sum of the three plane angles which form the solid angle at O is less than four right angles (Euclid, XI. 21); therefore

$$\frac{AB}{OA} + \frac{BC}{OA} + \frac{CA}{OA} \text{ is less than } 2\pi,$$

therefore, $AB + BC + CD$ is less than $2\pi \times OA$;
that is, the sum of the arcs is less than the circumference of a great circle.

31. The propositions contained in the preceding two Articles may be extended. Thus, if there be any polygon which has each of its angles less than two right angles, *any one side is less than the sum of all the others.* This may be proved by repeated use of Art. 29. Suppose, for example, that the figure has four sides, and let the angular points be denoted by A, B, C, D. Then

$$AD + BC \text{ is greater than } AC;$$

therefore, $AB + BC + CD$ is greater than $AC + CD$,
and *à fortiori* greater than AD.

Again, if there be any polygon which has each of its angles less than two right angles, *the sum of its sides will be less than the circumference of a great circle.* This follows from Euclid, XI. 21, in the manner shewn in Art. 30.

32. *The three angles of a spherical triangle are together greater than two right angles and less than six right angles.*

Let A, B, C be the *angles* of a spherical triangle; let a', b', c' be the *sides* of the polar triangle. Then by Art. 30,

$$a' + b' + c' \text{ is less than } 2\pi,$$

that is, $\pi - A + \pi - B + \pi - C$ is less than 2π;
therefore, $A + B + C$ is greater than π.

And since each of the angles A, B, C is less than π, the sum $A + B + C$ is less than 3π.

33. *The angles at the base of an isosceles spherical triangle are equal.*

Let ABC be a spherical triangle having $AC = BC$; let O be the centre of the sphere. Draw tangents at the points A and B to the arcs AC and BC respectively; these will meet OC produced at the same point S, and AS will be equal to BS.

Draw tangents AT, BT at the points A, B to the arc AB; then $AT = TB$;

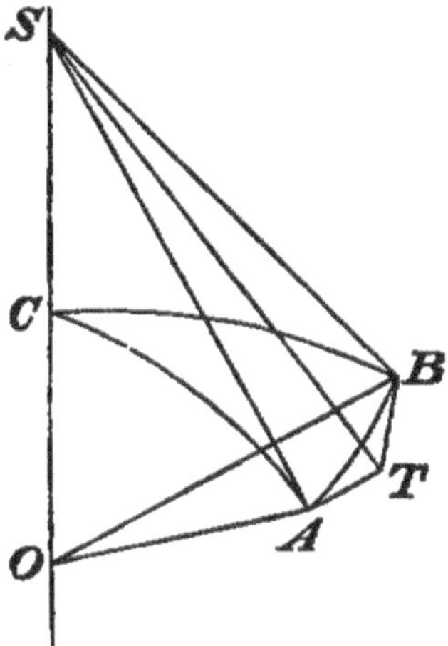

join TS. In the two triangles SAT, SBT the sides SA, AT, TS are equal to SB, BT, TS respectively; therefore the angle SAT is equal to the angle SBT; and these are the angles at the base of the spherical triangle.

The figure supposes AC and BC to be less than quadrants; if they are greater than quadrants the tangents to AC and BC will meet on CO produced through O instead of through C, and the demonstration may be completed as before. If AC and BC are quadrants, the angles at the base are right angles by Arts. 11 and 9.

34. *If two angles of a spherical triangle are equal, the opposite sides are equal.*

Since the primitive triangle has two equal angles, the polar triangle has two equal sides; therefore in the polar triangle the angles opposite the equal sides are equal by Art. 33. Hence in the primitive triangle the sides opposite the equal angles are equal.

35. *If one angle of a spherical triangle be greater than another, the side opposite the greater angle is greater than the side opposite the less angle.*

Let ABC be a spherical triangle, and let the angle ABC be greater than the angle BAC: then the side AC will be greater than the side BC. At B make the angle ABD equal to the angle BAD; then BD is equal to AD (Art. 34), and $BD+DC$ is greater than BC (Art. 29); therefore $AD+DC$ is greater than

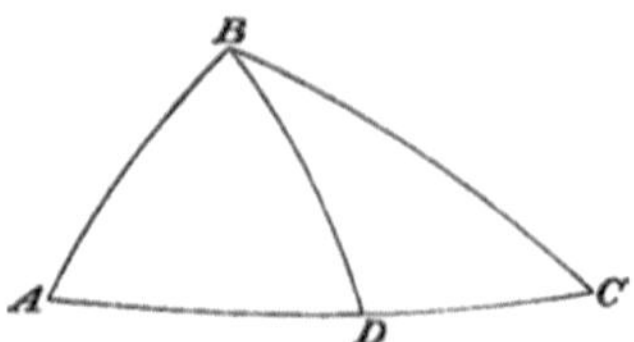

BC; that is, AC is greater than BC.

36. *If one side of a spherical triangle be greater than another, the angle opposite the greater side is greater than the angle opposite the less side.*

This follows from the preceding Article by means of the polar triangle.

Or thus; suppose the side AC greater than the side BC, then the angle ABC will be greater than the angle BAC. For the angle ABC cannot be less than the angle BAC by Art. 35, and the angle ABC cannot be equal to the angle BAC by Art. 34; therefore the angle ABC must be greater than the angle BAC.

This Chapter might be extended; but it is unnecessary to do so because the Trigonometrical formulæ of the next Chapter supply an easy method of investigating the theorems of Spherical Geometry. See Arts. 56, 57, and 58.

IV

Relations between the Trigonometrical Functions of the Sides and the Angles of a Spherical Triangle.

37. *To express the cosine of an angle of a triangle in terms of sines and cosines of the sides.*

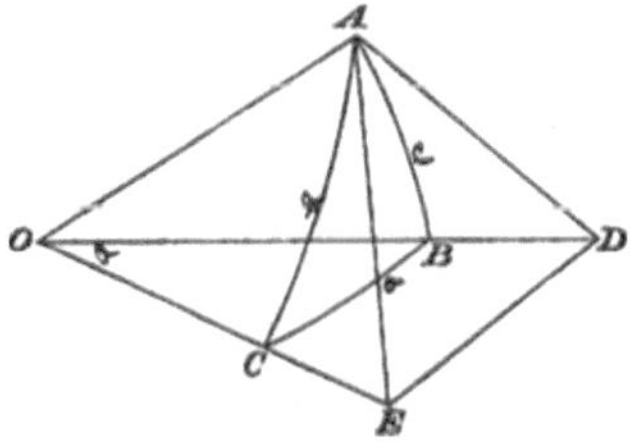

Let ABC be a spherical triangle, O the centre of the sphere. Let the tangent at A to the arc AC meet OC produced at E, and let the tangent at A to the arc AB meet OB produced at D; join ED. Thus the angle EAD is the angle A of the spherical triangle, and the angle EOD measures the side a.

From the triangles ADE and ODE we have

$$DE^2 = AD^2 + AE^2 - 2AD \cdot AE \cos A,$$
$$DE^2 = OD^2 + OE^2 - 2OD \cdot OE \cos a;$$

also the angles OAD and OAE are right angles, so that $OD^2 = OA^2 + AD^2$ and $OE^2 = OA^2 + AE^2$. Hence by subtraction we have

$$0 = 2OA^2 + 2AD \cdot AE \cos A - 2OD \cdot OE \cos a;$$

therefore $$\cos a = \frac{OA}{OE} \cdot \frac{OA}{OD} + \frac{AE}{OE} \cdot \frac{AD}{OD} \cos A;$$

that is $$\cos a = \cos b \cos c + \sin b \sin c \cos A.$$

Therefore $$\cos A = \frac{\cos a - \cos b \cos c}{\sin b \sin c}.$$

38. We have supposed, in the construction of the preceding Article, that the sides which contain the angle A are less than quadrants, for we have assumed that the tangents at A meet OB and OC respectively produced. We must now shew that the formulæ obtained is true when these sides are not less than quadrants. This we shall do by special examination of the cases in which one side or each side is greater than a quadrant or equal to a quadrant.

(1) Suppose only one of the sides which contain the angle A to be greater than a quadrant, for example, AB. Produce BA and BC to meet at B'; and put $AB' = c'$, $CB' = a'$.

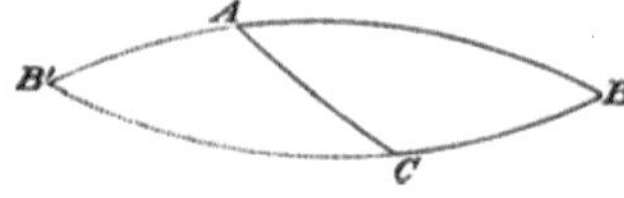

Then we have from the triangle $AB'C$, by what has been already proved,

$$\cos a' = \cos b \cos c' + \sin b \sin c' \cos B'AC;$$

but $a' = \pi - a$, $c' = \pi - c$, $B'AC = \pi - A$; thus

$$\cos a = \cos b \cos c + \sin b \sin c \cos A.$$

(2) Suppose both the sides which contain the angle A to be greater than quadrants. Produce AB and AC to meet at A'; put $A'B = c'$, $A'C = b'$; then from the triangle $A'BC$, as before,

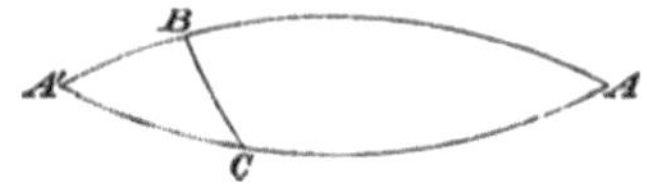

$$\cos a = \cos b' \cos c' + \sin b' \sin c' \cos A';$$

but $b' = \pi - b$, $c' = \pi - c$, $A' = A$; thus

$$\cos a = \cos b \cos c + \sin b \sin c \cos A.$$

(3) Suppose that one of the sides which contain the angle A is a quadrant, for example, AB; on AC, produced if necessary, take AD equal to a quadrant

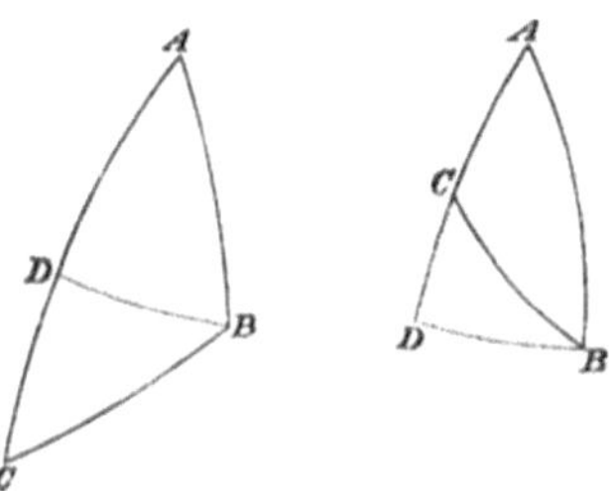

and draw BD. If BD is a quadrant B is a pole of AC (Art. 11); in this case $a = \frac{\pi}{2}$ and $A = \frac{\pi}{2}$ as well as $c = \frac{\pi}{2}$. Thus the formula to be verified reduces to the identity $0 = 0$. If BD be not a quadrant, the triangle BDC gives

$$\cos a = \cos CD \cos BD + \sin CD \sin BD \cos CDB,$$

and $$\cos CDB = 0, \quad \cos CD = \cos\left(\frac{\pi}{2} - b\right) = \sin b, \quad \cos BD = \cos A;$$

thus $$\cos a = \sin b \cos A;$$

and this is what the formula in Art. 37 becomes when $c = \frac{\pi}{2}$.

(4) Suppose that both the sides which contain the angle A are quadrants. The formula then becomes $\cos a = \cos A$; and this is obviously true, for A is now the pole of BC, and thus $A = a$.

Thus the formula in Art. 37 is proved to be universally true.

39. The formula in Art. 37 may be applied to express the cosine of any angle of a triangle in terms of sines and cosines of the sides; thus we have the three formulæ,

$$\cos a = \cos b \cos c + \sin b \sin c \cos A,$$
$$\cos b = \cos c \cos a + \sin c \sin a \cos B,$$
$$\cos c = \cos a \cos b + \sin a \sin b \cos C.$$

These may be considered as the fundamental equations of Spherical Trigonometry; we shall proceed to deduce various formulæ from them.

40. *To express the sine of an angle of a spherical triangle in terms of trigonometrical functions of the sides.*

We have $\cos A = \dfrac{\cos a - \cos b \cos c}{\sin b \sin c};$

therefore
$$\sin^2 A = 1 - \left(\frac{\cos a - \cos b \cos c}{\sin b \sin c}\right)^2$$
$$= \frac{(1-\cos^2 b)(1-\cos^2 c) - (\cos a - \cos b \cos c)^2}{\sin^2 b \sin^2 c}$$
$$= \frac{1 - \cos^2 a - \cos^2 b - \cos^2 c + 2\cos a \cos b \cos c}{\sin^2 b \sin^2 c};$$

therefore
$$\sin A = \frac{\sqrt{(1 - \cos^2 a - \cos^2 b - \cos^2 c + 2\cos a \cos b \cos c)}}{\sin b \sin c}.$$

The radical on the right-hand side must be taken with the positive sign, because $\sin b$, $\sin c$, and $\sin A$ are all positive.

41. From the value of $\sin A$ in the preceding Article it follows that

$$\frac{\sin A}{\sin a} = \frac{\sin B}{\sin b} = \frac{\sin C}{\sin c},$$

for each of these is equal to the same expression, namely,

$$\frac{\sqrt{(1 - \cos^2 a - \cos^2 b - \cos^2 c + 2 \cos a \cos b \cos c)}}{\sin a \sin b \sin c}.$$

Thus *the sines of the angles of a spherical triangle are proportional to the sines of the opposite sides.* We will give an independent proof of this proposition in the following Article.

42. *The sines of the angles of a spherical triangle are proportional to the sines of the opposite sides.*

Let ABC be a spherical triangle, O the centre of the sphere. Take any point P in OA, draw PD perpendicular to the plane BOC, and from D draw DE,

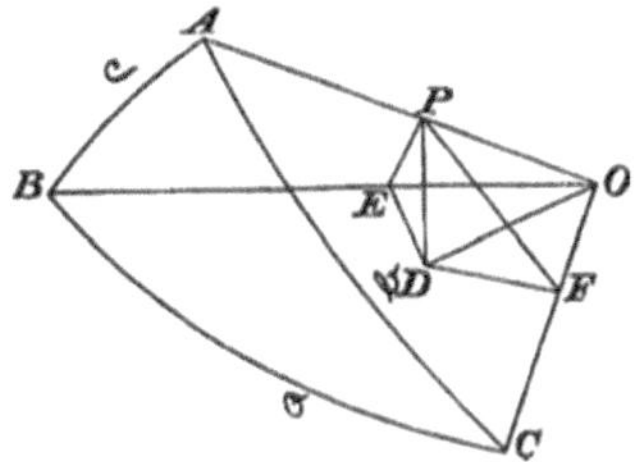

DF perpendicular to OB, OC respectively; join PE, PF, OD.

Since PD is perpendicular to the plane BOC, it makes right angles with every straight line meeting it in that plane; hence

$$PE^2 = PD^2 + DE^2 = PO^2 - OD^2 + DE^2 = PO^2 - OE^2;$$

thus PEO is a right angle. Therefore $PE = OP \sin POE = OP \sin c$; and $PD = PE \sin PED = PE \sin B = OP \sin c \sin B$.

Similarly, $PD = OP \sin b \sin C$; therefore

$$OP \sin c \sin B = OP \sin b \sin C;$$

therefore $$\frac{\sin B}{\sin C} = \frac{\sin b}{\sin c}.$$

The figure supposes b, c, B, and C each less than a right angle; it will be found on examination that the proof will hold when the figure is modified to meet any case which can occur. If, for instance, B alone is greater than a right angle, the point D will fall beyond OB instead of between OB and OC; then PED will be the *supplement* of B, and thus $\sin PED$ is still equal to $\sin B$.

43. *To shew that* $\cot a \sin b = \cot A \sin C + \cos b \cos C$.

We have
$$\cos a = \cos b \cos c + \sin b \sin c \cos A,$$
$$\cos c = \cos a \cos b + \sin a \sin b \cos C,$$
$$\sin c = \sin a \frac{\sin C}{\sin A}.$$

Substitute the values of $\cos c$ and $\sin c$ in the first equation; thus

$$\cos a = (\cos a \cos b + \sin a \sin b \cos C) \cos b + \frac{\sin a \sin b \cos A \sin C}{\sin A};$$

by transposition

$$\cos a \sin^2 b = \sin a \sin b \cos b \cos C + \sin a \sin b \cot A \sin C;$$

divide by $\sin a \sin b$; thus

$$\cot a \sin b = \cos b \cos C + \cot A \sin C.$$

44. By interchanging the letters five other formulæ may be obtained like that in the preceding Article; the whole six formulæ will be as follows:

$$\cot a \sin b = \cot A \sin C + \cos b \cos C,$$
$$\cot b \sin a = \cot B \sin C + \cos a \cos C,$$
$$\cot b \sin c = \cot B \sin A + \cos c \cos A,$$
$$\cot c \sin b = \cot C \sin A + \cos b \cos A,$$
$$\cot c \sin a = \cot C \sin B + \cos a \cos B,$$
$$\cot a \sin c = \cot A \sin B + \cos c \cos B.$$

45. *To express the sine, cosine, and tangent, of half an angle of a triangle as functions of the sides.*

We have, by Art. 37, $\cos A = \dfrac{\cos a - \cos b \cos c}{\sin b \sin c}$;

therefore $\quad 1-\cos A = 1 - \dfrac{\cos a - \cos b\cos c}{\sin b\sin c} = \dfrac{\cos(b-c)-\cos a}{\sin b\sin c}$;

therefore $$\sin^2\frac{A}{2} = \frac{\sin\frac{1}{2}(a+b-c)\sin\frac{1}{2}(a-b+c)}{\sin b\sin c}.$$

Let $2s = a+b+c$, so that s is half the sum of the sides of the triangle; then

$$a+b-c = 2s-2c = 2(s-c), \quad a-b+c = 2s-2b = 2(s-b);$$

thus, $$\sin^2\frac{A}{2} = \frac{\sin(s-b)\sin(s-c)}{\sin b\sin c},$$

and $$\sin\frac{A}{2} = \sqrt{\left\{\frac{\sin(s-b)\sin(s-c)}{\sin b\sin c}\right\}}$$

Also, $\quad 1+\cos A = 1+\dfrac{\cos a - \cos b\cos c}{\sin b\sin c} = \dfrac{\cos a - \cos(b+c)}{\sin b\sin c}$;

therefore

$$\cos^2\frac{A}{2} = \frac{\sin\frac{1}{2}(a+b+c)\sin\frac{1}{2}(b+c-a)}{\sin b\sin c} = \frac{\sin s\sin(s-a)}{\sin b\sin c},$$

and $$\cos\frac{A}{2} = \sqrt{\left\{\frac{\sin s\sin(s-a)}{\sin b\sin c}\right\}}$$

From the expressions for $\sin\dfrac{A}{2}$ and $\cos\dfrac{A}{2}$ we deduce

$$\tan\frac{A}{2} = \sqrt{\left\{\frac{\sin(s-b)\sin(s-c)}{\sin s\sin(s-a)}\right\}}.$$

The positive sign must be given to the radicals which occur in this Article, because $\dfrac{A}{2}$ is less than a right angle, and therefore its sine, cosine, and tangent are all positive.

46. Since $\sin A = 2\sin\dfrac{A}{2}\cos\dfrac{A}{2}$, we obtain

$$\sin A = \frac{2}{\sin b\sin c}\{\sin s\sin(s-a)\sin(s-b)\sin(s-c)\}^{\frac{1}{2}}.$$

It may be shewn that the expression for $\sin A$ in Art. 40 agrees with the present expression by putting the numerator of that expression in factors, as in

Plane Trigonometry, Art. 115. We shall find it convenient to use a symbol for the radical in the value of $\sin A$; we shall denote it by n, so that

$$n^2 = \sin s \sin(s-a)\sin(s-b)\sin(s-c),$$

and
$$4n^2 = 1 - \cos^2 a - \cos^2 b - \cos^2 c + 2\cos a \cos b \cos c.$$

47. *To express the cosine of a side of a triangle in terms of sines and cosines of the angles.*

In the formula of Art. 37 we may, by Art. 28, change the sides into the supplements of the corresponding angles and the angle into the supplement of the corresponding side; thus

$$\cos(\pi - A) = \cos(\pi - B)\cos(\pi - C) + \sin(\pi - B)\sin(\pi - C)\cos(\pi - a),$$

that is,
$$\cos A = -\cos B \cos C + \sin B \sin C \cos a.$$

Similarly
$$\cos B = -\cos C \cos A + \sin C \sin A \cos b,$$

and
$$\cos C = -\cos A \cos B + \sin A \sin B \cos c.$$

48. The formulæ in Art. 44 will of course remain true when the angles and sides are changed into the supplements of the corresponding sides and angles respectively; it will be found, however, that no *new* formulæ are thus obtained, but only the *same* formulæ over again. This consideration will furnish some assistance in retaining those formulæ accurately in the memory.

49. *To express the sine, cosine, and tangent, of half a side of a triangle as functions of the angles.*

We have, by Art. 47, $\cos a = \dfrac{\cos A + \cos B \cos C}{\sin B \sin C}$;

therefore

$$1 - \cos a = 1 - \frac{\cos A + \cos B \cos C}{\sin B \sin C} = -\frac{\cos A + \cos(B+C)}{\sin B \sin C};$$

therefore
$$\sin^2 \frac{a}{2} = -\frac{\cos \frac{1}{2}(A+B+C)\cos\frac{1}{2}(B+C-A)}{\sin B \sin C}.$$

Let $2S = A + B + C$; then $B + C - A = 2(S - A)$, therefore

$$\sin^2 \frac{a}{2} = -\frac{\cos S \cos(S-A)}{\sin B \sin C},$$

and $$\sin\frac{a}{2} = \sqrt{\left\{-\frac{\cos S\cos(S-A)}{\sin B\sin C}\right\}}.$$

Also $$1+\cos a = 1+\frac{\cos A+\cos B\cos C}{\sin B\sin C} = \frac{\cos A+\cos(B-C)}{\sin B\sin C};$$

therefore

$$\cos^2\frac{a}{2} = \frac{\cos\frac{1}{2}(A-B+C)\cos\frac{1}{2}(A+B-C)}{\sin B\sin C} = \frac{\cos(S-B)\cos(S-C)}{\sin B\sin C},$$

and $$\cos\frac{a}{2} = \sqrt{\left\{\frac{\cos(S-B)\cos(S-C)}{\sin B\sin C}\right\}}.$$

Hence $$\tan\frac{a}{2} = \sqrt{\left\{-\frac{\cos S\cos(S-A)}{\cos(S-B)\cos(S-C)}\right\}}.$$

The positive sign must be given to the radicals which occur in this Article, because $\frac{a}{2}$ is less than a right angle.

50. The expressions in the preceding Article may also be obtained immediately from those given in Art. 45 by means of Art. 28.

It may be remarked that the values of $\sin\frac{a}{2}$, $\cos\frac{a}{2}$, and $\tan\frac{a}{2}$ are *real*. For S is greater than one right angle and less than three right angles by Art. 32; therefore $\cos S$ is *negative*. And in the polar triangle any side is less than the sum of the other two; thus $\pi-A$ is less than $\pi-B+\pi-C$; therefore $B+C-A$ is less than π; therefore $S-A$ is less than $\frac{\pi}{2}$, and $B+C-A$ is algebraically greater than $-\pi$, so that $S-A$ is algebraically greater than $-\frac{\pi}{2}$; therefore $\cos(S-A)$ is *positive*. Similarly also $\cos(S-B)$ and $\cos(S-C)$ are positive. Hence the values of $\sin\frac{a}{2}$, $\cos\frac{a}{2}$, and $\tan\frac{a}{2}$ are real.

51. Since $\sin a = 2\sin\frac{a}{2}\cos\frac{a}{2}$, we obtain

$$\sin a = \frac{2}{\sin B\sin C}\{-\cos S\cos(S-A)\cos(S-B)\cos(S-C)\}^{\frac{1}{2}}.$$

We shall use N for $\{-\cos S\cos(S-A)\cos(S-B)\cos(S-C)\}^{\frac{1}{2}}$.

52. *To demonstrate Napier's Analogies.*

We have $$\frac{\sin A}{\sin a} = \frac{\sin B}{\sin b} = m \text{ suppose;}$$

then, by a theorem of Algebra,

$$m = \frac{\sin A + \sin B}{\sin a + \sin b}, \tag{1}$$

and also $$m = \frac{\sin A - \sin B}{\sin a - \sin b}. \tag{2}$$

Now $$\cos A + \cos B \cos C = \sin B \sin C \cos a = m \sin C \sin b \cos a,$$

and $$\cos B + \cos A \cos C = \sin A \sin C \cos b = m \sin C \sin a \cos b,$$

therefore, by addition,

$$(\cos A + \cos B)(1 + \cos C) = m \sin C \sin(a + b); \tag{3}$$

therefore by (1) we have

$$\frac{\sin A + \sin B}{\cos A + \cos B} = \frac{\sin a + \sin b}{\sin(a + b)} \frac{1 + \cos C}{\sin C},$$

that is, $$\tan \tfrac{1}{2}(A + B) = \frac{\cos \frac{1}{2}(a - b)}{\cos \frac{1}{2}(a + b)} \cot \frac{C}{2}. \tag{4}$$

Similarly from (3) and (2) we have

$$\frac{\sin A - \sin B}{\cos A + \cos B} = \frac{\sin a - \sin b}{\sin(a + b)} \frac{1 + \cos C}{\sin C},$$

that is, $$\tan \frac{1}{2}(A - B) = \frac{\sin \frac{1}{2}(a - b)}{\sin \frac{1}{2}(a + b)} \cot \frac{C}{2}. \tag{5}$$

By writing $\pi - A$ for a, and so on in (4) and (5) we obtain

$$\tan \tfrac{1}{2}(a + b) = \frac{\cos \frac{1}{2}(A - B)}{\cos \frac{1}{2}(A + B)} \tan \frac{c}{2}, \tag{6}$$

$$\tan \tfrac{1}{2}(a - b) = \frac{\sin \frac{1}{2}(A - B)}{\sin \frac{1}{2}(A + B)} \tan \frac{c}{2}. \tag{7}$$

The formulæ (4), (5), (6), (7) may be put in the form of proportions or analogies, and are called from their discoverer *Napier's Analogies:* the last two may be demonstrated without recurring to the polar triangle by starting with the formulæ in Art. 39.

53. In equation (4) of the preceding Article, $\cos\frac{1}{2}(a-b)$ and $\cot\dfrac{C}{2}$ are necessarily positive quantities; hence the equation shews that $\tan\frac{1}{2}(A+B)$ and $\cos\frac{1}{2}(a+b)$ are of the same sign; thus $\frac{1}{2}(A+B)$ and $\frac{1}{2}(a+b)$ are either both less than a right angle or both greater than a right angle. This is expressed by saying that $\frac{1}{2}(A+B)$ and $\frac{1}{2}(a+b)$ *are of the same affection.*

54. *To demonstrate Delambre's Analogies.*

We have $\cos c = \cos a\cos b + \sin a\sin b\cos C$; therefore

$$\begin{aligned}1+\cos c &= 1+\cos a\cos b+\sin a\sin b(\cos^2\tfrac{1}{2}C-\sin^2\tfrac{1}{2}C)\\ &= \{1+\cos(a-b)\}\cos^2\tfrac{1}{2}C+\{1+\cos(a+b)\}\sin^2\tfrac{1}{2}C;\end{aligned}$$

therefore $\cos^2\frac{1}{2}c=\cos^2\frac{1}{2}(a-b)\cos^2\frac{1}{2}C+\cos^2\frac{1}{2}(a+b)\sin^2\frac{1}{2}C$.
Similarly, $\sin^2\frac{1}{2}c=\sin^2\frac{1}{2}(a-b)\cos^2\frac{1}{2}C+\sin^2\frac{1}{2}(a+b)\sin^2\frac{1}{2}C$.

Now add unity to the square of each member of Napier's first two analogies; hence by the formulæ just proved

$$\sec^2\tfrac{1}{2}(A+B)=\frac{\cos^2\frac{1}{2}c}{\cos^2\frac{1}{2}(a+b)\sin^2\frac{1}{2}C},$$

$$\sec^2\tfrac{1}{2}(A-B)=\frac{\sin^2\frac{1}{2}c}{\sin^2\frac{1}{2}(a+b)\sin^2\frac{1}{2}C}.$$

Extract the square roots; thus, since $\frac{1}{2}(A+B)$ and $\frac{1}{2}(a+b)$ are of the same affection, we obtain

$$\cos\tfrac{1}{2}(A+B)\cos\tfrac{1}{2}c=\cos\tfrac{1}{2}(a+b)\sin\tfrac{1}{2}C,\qquad(1)$$

$$\cos\tfrac{1}{2}(A-B)\sin\tfrac{1}{2}c=\sin\tfrac{1}{2}(a+b)\sin\tfrac{1}{2}C.\qquad(2)$$

Multiply the first two of Napier's analogies respectively by these results; thus

$$\sin\tfrac{1}{2}(A+B)\cos\tfrac{1}{2}c=\cos\tfrac{1}{2}(a-b)\cos\tfrac{1}{2}C,\qquad(3)$$

$$\sin\tfrac{1}{2}(A-B)\sin\tfrac{1}{2}c=\sin\tfrac{1}{2}(a-b)\cos\tfrac{1}{2}C.\qquad(4)$$

The last four formulæ are commonly, but improperly, called *Gauss's Theorems*; they were first given by Delambre in the *Connaissance des Tems* for 1809, page 445. See the *Philosophical Magazine* for February, 1873.

55. The properties of supplemental triangles were proved geometrically in Art. 27, and by means of these properties the formulæ in Art. 47 were obtained; but these formulæ may be deduced analytically from those in Art. 39, and thus the whole subject may be made to depend on the formulæ of Art. 39.

For from Art. 39 we obtain expressions for $\cos A$, $\cos B$, $\cos C$; and from these we find

$$\cos A + \cos B \cos C$$
$$= \frac{(\cos a - \cos b \cos c)\sin^2 a + (\cos b - \cos a \cos c)(\cos c - \cos a \cos b)}{\sin^2 a \sin b \sin c}.$$

In the numerator of this fraction write $1 - \cos^2 a$ for $\sin^2 a$; thus the numerator will be found to reduce to

$$\cos a(1 - \cos^2 a - \cos^2 b - \cos^2 c + 2\cos a \cos b \cos c),$$

and this is equal to $\cos a \sin B \sin C \sin^2 a \sin b \sin c$, (Art. 41);

therefore $$\cos A + \cos B \cos C - \cos a \sin B \sin C.$$

Similarly the other two corresponding formulæ may be proved.

Thus the formulæ in Art. 47 are established; and therefore, without assuming the existence and properties of the Polar Triangle, we deduce the following theorem: *If the sides and angles of a spherical triangle be changed respectively into the supplements of the corresponding angles and sides, the fundamental formulæ of Art. 39 hold good, and therefore also all results deducible from them.*

56. The formulæ in the present Chapter may be applied to establish analytically various propositions respecting spherical triangles which either have been proved geometrically in the preceding Chapter, or may be so proved. Thus, for example, the second of Napier's analogies is

$$\tan \tfrac{1}{2}(A - B) = \frac{\sin \frac{1}{2}(a - b)}{\sin \frac{1}{2}(a + b)} \cot \frac{C}{2};$$

this shews that $\frac{1}{2}(A - B)$ is positive, negative, or zero, according as $\frac{1}{2}(a - b)$ is positive, negative, or zero; thus we obtain all the results included in Arts. 33...36.

57. *If two triangles have two sides of the one equal to two sides of the other, each to each, and likewise the included angles equal, then their other angles will be equal, each to each, and likewise their bases will be equal.*

We may shew that the bases are equal by applying the first formula in Art. 39 to each triangle, supposing b, c, and A the same in the two triangles; then the remaining two formulæ of Art. 39 will shew that B and C are the same in the two triangles.

It should be observed that the two triangles in this case are *not* necessarily such that one may be made to *coincide with the other by superposition.* The sides of one may be equal to those of the other, each to each, but in a reverse order, as in the following figures.

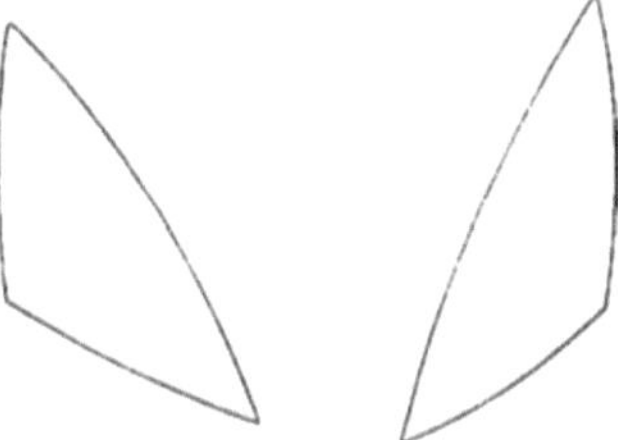

Two triangles which are equal in this manner are said to be *symmetrically* equal; when they are equal so as to admit of superposition they are said to be *absolutely* equal.

58. *If two spherical triangles have two sides of the one equal to two sides of the other, each to each, but the angle which is contained by the two sides of the one greater than the angle which is contained by the two sides which are equal to them of the other, the base of that which has the greater angle will be greater than the base of the other; and conversely.*

Let b and c denote the sides which are equal in the two triangles; let a be the base and A the opposite angle of one triangle, and a' and A' similar quantities for the other. Then

$$\cos a = \cos b \cos c + \sin b \sin c \cos A,$$
$$\cos a' = \cos b \cos c + \sin b \sin c \cos A';$$

therefore $$\cos a - \cos a' + \sin b \sin c(\cos A - \cos A');$$

that is,

$$\sin \tfrac{1}{2}(a + a') \sin \tfrac{1}{2}(a - a') = \sin b \sin c \sin \tfrac{1}{2}(A + A') \sin \tfrac{1}{2}(A - A');$$

this shews that $\frac{1}{2}(a - a')$ and $\frac{1}{2}(A - A')$ are of the same sign.

59. *If on a sphere any point be taken within a circle which is not its pole, of all the arcs which can be drawn from that point to the circumference, the greatest is that in which the pole is, and the other part of that produced is the least; and of any others, that which is nearer to the greatest is always greater than one more remote; and from the same point to the circumference there can be drawn only two arcs which are equal to each other, and these make equal angles with the shortest arc on opposite sides of it.*

This follows readily from the preceding three Articles.

60. We will give another proof of the fundamental formulæ in Art. 39, which is very simple, requiring only a knowledge of the elements of Co-ordinate Geometry.

Suppose ABC any spherical triangle, O the centre of the sphere, take O as the origin of co-ordinates, and let the axis of z pass through C. Let x_1, y_1, z_1 be the co-ordinates of A, and x_2, y_2, z_2 those of B; let r be the radius of the sphere. Then the square on the straight line AB is equal to

$$(x_1 - x_2)^2 + (y_1 - y_2)^2 + (z_1 - z_2)^2,$$

and also to $$r^2 + r^2 - 2r^2 \cos AOB;$$

and $x_1^2 + y_1^2 + z_1^2 = r^2$, $x_2^2 + y_2^2 + z_2^2 = r^2$, thus

$$x_1x_2 + y_1y_2 + z_1z_2 = r^2 \cos AOB.$$

Now make the usual substitutions in passing from rectangular to polar co-ordinates, namely,

$$z_1 = r\cos\theta_1, \qquad x_1 = r\sin\theta_1\cos\phi_1, \qquad y_1 = r\sin\theta_1\sin\phi_1,$$

$$z_2 = r\cos\theta_2, \qquad x_2 = r\sin\theta_2\cos\phi_2, \qquad y_2 = r\sin\theta_2\sin\phi_2;$$

thus we obtain

$$\cos\theta_2\cos\theta_1 + \sin\theta_2\sin\theta_1\cos(\phi_1 - \phi_2) = \cos AOB,$$

that is, in the ordinary notation of Spherical Trigonometry,

$$\cos a\cos b + \sin a\sin b\cos C = \cos c.$$

This method has the advantage of giving a *perfectly general proof*, as all the equations used are universally true.

EXAMPLES.

1. If $A = a$, shew that B and b are equal or supplemental, as also C and c.

2. If one angle of a triangle be equal to the sum of the other two, the greatest side is double of the distance of its middle point from the opposite angle.

3. When does the polar triangle coincide with the primitive triangle?

4. If D be the middle point of AB, shew that

$$\cos AC + \cos BC = 2\cos\tfrac{1}{2}AB\cos CD.$$

5. If two angles of a spherical triangle be respectively equal to the sides opposite to them, shew that the remaining side is the supplement of the remaining angle; or else that the triangle has two quadrants and two right angles, and then the remaining side is equal to the remaining angle.

6. In an equilateral triangle, shew that $2\cos\dfrac{a}{2}\sin\dfrac{A}{2} = 1$.

7. In an equilateral triangle, shew that $\tan^2\dfrac{a}{2} = 1 - 2\cos A$; hence deduce the limits between which the sides and the angles of an equilateral triangle are restricted.

8. In an equilateral triangle, shew that $\sec A = 1 + \sec a$.

9. If the three sides of a spherical triangle be halved and a new triangle formed, the angle θ between the new sides $\frac{b}{2}$ and $\frac{c}{2}$ is given by $\cos\theta = \cos A + \frac{1}{2}\tan\frac{b}{2}\tan\frac{c}{2}\sin^2\theta$.

10. AB, CD are quadrants on the surface of a sphere intersecting at E, the extremities being joined by great circles: shew that

$$\cos AEC = \cos AC \cos BD - \cos BC \cos AD.$$

11. If $b + c = \pi$, shew that $\sin 2B + \sin 2C = 0$.

12. If DE be an arc of a great circle bisecting the sides AB, AC of a spherical triangle at D and E, P a pole of DE, and PB, PD, PE, PC be joined by arcs of great circles, shew that the angle BPC = twice the angle DPE.

13. In a spherical triangle shew that

$$\sin b \sin c + \cos b \cos c \cos A = \sin B \sin C - \cos B \cos C \cos a.$$

14. If D be any point in the side BC of a triangle, shew that

$$\cos AD \sin BC = \cos AB \sin DC + \cos AC \sin BD.$$

15. In a spherical triangle shew that θ, ϕ, ψ be the lengths of arcs of great circles drawn from A, B, C perpendicular to the opposite sides,

$$\begin{aligned}\sin a \sin\theta &= \sin b \sin\phi = \sin c \sin\psi\\ &= \sqrt{(1 - \cos^2 a - \cos^2 b - \cos^2 c + 2\cos a \cos b \cos c)}.\end{aligned}$$

16. In a spherical triangle, if θ, ϕ, ψ be the arcs bisecting the angles A, B, C respectively and terminated by the opposite sides, shew that

$$\cot\theta\cos\frac{A}{2} + \cot\phi\cos\frac{B}{2} + \cot\psi\cos\frac{C}{2} = \cot a + \cot b + \cot c.$$

17. Two ports are in the same parallel of latitude, their common latitude being l and their difference of longitude 2λ: shew that the saving of distance in sailing from one to the other on the great circle, instead of sailing due East or

West, is

$$2r\{\lambda \cos l - \sin^{-1}(\sin \lambda \cos l)\},$$

λ being expressed in circular measure, and r being the radius of the Earth.

18. If a ship be proceeding uniformly along a great circle and the observed latitudes be l_1, l_2, l_3, at equal intervals of time, in each of which the distance traversed is s, shew that

$$s = r \cos^{-1} \frac{\sin \frac{1}{2}(l_1 + l_3) \cos \frac{1}{2}(l_1 - l_3)}{\sin l_2},$$

r denoting the Earth's radius: and shew that the change of longitude may also be found in terms of the three latitudes.

V

SOLUTION OF RIGHT-ANGLED TRIANGLES.

61. In every spherical triangle there are six elements, namely, the three sides and the three angles, besides the radius of the sphere, which is supposed constant. The solution of spherical triangles is the process by which, when the values of a sufficient number of the six elements are given, we calculate the values of the remaining elements. It will appear, as we proceed, that when the values of three of the elements are given, those of the remaining three can generally be found. We begin with the right-angled triangle: here two elements, in addition to the right angle, will be supposed known.

62. The formulæ requisite for the solution of right-angled triangles may be obtained from the preceding Chapter by supposing one of the angles a right angle, as C for example. They may also be obtained very easily in an independent manner, as we will now shew.

Let ABC be a spherical triangle having a right angle at C; let O be the centre of the sphere. From any point P in OA draw PM perpendicular to OC, and from M draw MN perpendicular to OB, and join PN. Then PM is perpendicular to MN, because the plane AOC is perpendicular to the plane

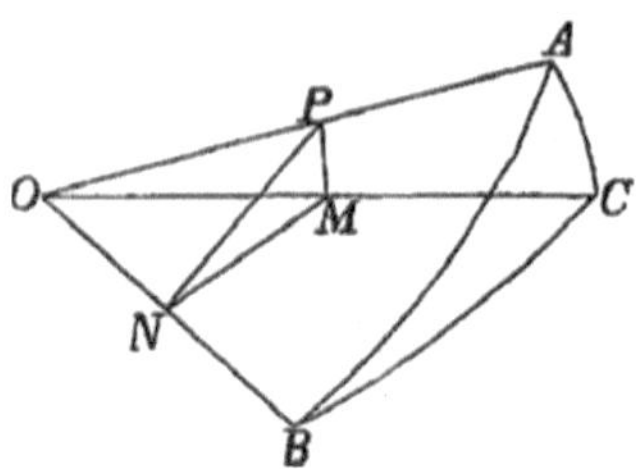

BOC; hence

$$PN^2 = PM^2 + MN^2 = OP^2 - OM^2 + OM^2 - ON^2 = OP^2 - ON^2;$$

therefore PNO is a right angle. And

$$\frac{ON}{OP} = \frac{ON}{OM} \cdot \frac{OM}{OP}, \quad \text{that is, } \cos c = \cos a \cos b, \tag{1}$$

$$\left.\begin{aligned} \frac{PM}{OP} = \frac{PM}{PN} \cdot \frac{PN}{OP}, \quad \text{that is, } \sin b &= \sin B \sin c \\ \text{Similarly,} \quad \sin a &= \sin A \sin c \end{aligned}\right\}, \tag{2}$$

$$\left.\begin{aligned} \frac{MN}{ON} = \frac{MN}{PN} \cdot \frac{PN}{ON}, \quad \text{that is, } \tan a &= \cos B \tan c \\ \text{Similarly,} \quad \tan b &= \cos A \tan c \end{aligned}\right\}, \tag{3}$$

$$\left.\begin{aligned} \frac{PM}{OM} = \frac{PM}{MN} \cdot \frac{MN}{OM}, \quad \text{that is, } \tan b &= \tan B \sin a \\ \text{Similarly,} \quad \tan a &= \tan A \sin b \end{aligned}\right\}. \tag{4}$$

Multiply together the two formulæ (4); thus,

$$\tan A \tan B = \frac{\tan a \tan b}{\sin a \sin b} = \frac{1}{\cos a \cos b} = \frac{1}{\cos c} \text{ by (1);}$$

therefore $$\cos c = \cot A \cot B. \tag{5}$$

Multiply crosswise the second formula in (2) and the first in (3); thus $\sin a \cos B \tan c = \tan a \sin A \sin c$;

therefore $$\cos B = \frac{\sin A \cos c}{\cos a} = \sin A \cos b \text{ by (1).}$$

Thus $\qquad\qquad \left.\begin{aligned}\cos B &= \sin A \cos b\\ \cos A &= \sin B \cos a\end{aligned}\right\}. \qquad (6)$
Similarly

These six formulæ comprise ten equations; and thus we can solve every case of right-angled triangles. For every one of these ten equations is a distinct combination involving three out of the five quantities a, b, c, A, B; and out of five quantities only ten combinations of three can be formed. Thus any two of the five quantities being given and a third required, some one of the preceding ten equations will serve to determine that third quantity.

63. As we have stated, the above six formulæ may be obtained from those given in the preceding Chapter by supposing C a right angle. Thus (1) follows from Art. 39, (2) from Art. 41, (3) from the fourth and fifth equations of Art. 44, (4) from the first and second equations of Art. 44, (5) from the third equation of Art. 47, (6) from the first and second equations of Art. 47.

Since the six formulæ may be obtained from those given in the preceding Chapter which have been proved to be universally true, we do not stop to shew that the demonstration of Art. 62 may be applied to every case which can occur; the student may for exercise investigate the modifications which will be necessary when we suppose one or more of the quantities a, b, c, A, B equal to a right angle or greater than a right angle.

64. Certain properties of right-angled triangles are deducible from the formulæ of Art. 62.

From (1) it follows that $\cos c$ has the same sign as the product $\cos a \cos b$; hence either all the cosines are positive, or else only one is positive. Therefore *in a right-angled triangle either all the three sides are less than quadrants, or else one side is less than a quadrant and the other two sides are greater than quadrants.*

From (4) it follows that $\tan a$ has the same sign as $\tan A$. Therefore A and a are either both greater than $\frac{\pi}{2}$, or both less than $\frac{\pi}{2}$; this is expressed by saying that A and a are of the *same affection.* Similarly B and b are of the same affection.

65. The formulæ of Art. 62 are comprised in the following enunciations, which the student will find it useful to remember; the results are distinguished by the same numbers as have been already applied to them in Art. 62; the side opposite the right angle is called the *hypotenuse:*

Cos hyp = product of cosines of sides .... (1),

Cos hyp = product of cotangents of angles .... (5),

Sine side = sine of opposite angle × sine hyp .... (2),

Tan side = tan hyp × cos included angle .... (3),

Tan side = tan opposite angle × sine of other side ... (4),

Cos angle = cos opposite side × sine of other angle ... (6).

66. *Napier's Rules.* The formulæ of Art. 62 are comprised in two rules, which are called, from their inventor, *Napier's Rules of Circular Parts.* Napier was also the inventor of Logarithms, and the Rules of Circular Parts were first published by him in a work entitled *Mirifici Logarithmorum Canonis Descriptio......* Edinburgh, 1614. These rules we will now explain.

The right angle is left out of consideration; the two sides which include the right angle, the complement of the hypotenuse, and the complements of the other angles are called the *circular parts* of the triangle. Thus there are *five*

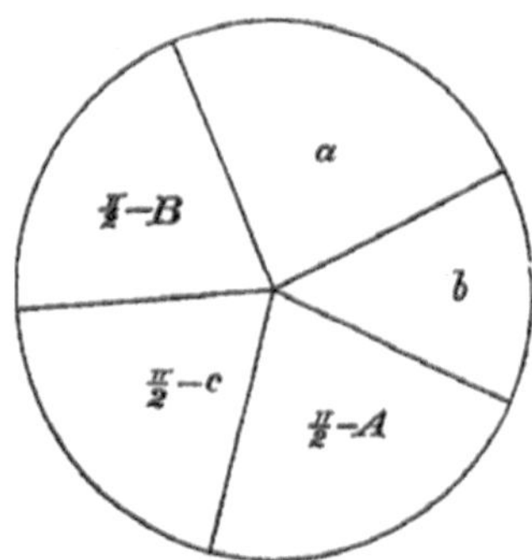

circular parts, namely, a, b, $\frac{\pi}{2}-A$, $\frac{\pi}{2}-c$, $\frac{\pi}{2}-B$; and these are supposed to be ranged round a circle in the order in which they naturally occur with respect to the triangle.

Any one of the five parts may be selected and called the *middle part,* then the two parts next to it are called *adjacent parts,* and the remaining two parts are called *opposite parts.* For example, if $\frac{\pi}{2}-B$ is selected as the middle part, then the adjacent parts are a and $\frac{\pi}{2}-c$, and the opposite parts are b and $\frac{\pi}{2}-A$.

Then Napier's Rules are the following:

sine of the middle part = product of tangents of adjacent parts,

sine of the middle part = product of cosines of opposite parts.

67. Napier's Rules may be demonstrated by shewing that they agree with the results already established. The following table shews the required agreement: in the first column are given the *middle parts*, in the second column the results of Napier's Rules, and in the third column the same results expressed as in Art. 62, with the number for reference used in that Article.

$\frac{\pi}{2}-c$	$\sin\left(\frac{\pi}{2}-c\right)=\tan\left(\frac{\pi}{2}-A\right)\tan\left(\frac{\pi}{2}-B\right)$	$\cos c=\cot A\cot B\ldots(5)$,
	$\sin\left(\frac{\pi}{2}-c\right)=\cos a\cos b$	$\cos c=\cos a\cos b\ldots.(1)$,
$\frac{\pi}{2}-B$	$\sin\left(\frac{\pi}{2}-B\right)=\tan a\tan\left(\frac{\pi}{2}-c\right)$	$\cos B=\tan a\cot c\ldots.(3)$,
	$\sin\left(\frac{\pi}{2}-B\right)=\cos b\cos\left(\frac{\pi}{2}-A\right)$	$\cos B=\cos b\sin A\ldots.(6)$,
a	$\sin a=\tan b\tan\left(\frac{\pi}{2}-B\right)$	$\sin a=\tan b\cot B\ldots.(4)$,
	$\sin a=\cos\left(\frac{\pi}{2}-A\right)\cos\left(\frac{\pi}{2}-c\right)$	$\sin a=\sin A\sin c\ldots.(2)$,
b	$\sin b=\tan\left(\frac{\pi}{2}-A\right)\tan a$	$\sin b=\cot A\tan a\ldots(4)$,
	$\sin b=\cos\left(\frac{\pi}{2}-B\right)\cos\left(\frac{\pi}{2}-c\right)$	$\sin b=\sin B\sin c\ldots.(2)$,
$\frac{\pi}{2}-A$	$\sin\left(\frac{\pi}{2}-A\right)=\tan b\tan\left(\frac{\pi}{2}-c\right)$	$\cos A=\tan b\cot c\ldots.(3)$,
	$\sin\left(\frac{\pi}{2}-A\right)=\cos a\cos\left(\frac{\pi}{2}-B\right)$	$\cos A=\cos a\sin B\ldots.(6)$.

The last four cases need not have been given, since it is obvious that they are only repetitions of what had previously been given; the seventh and eighth are repetitions of the fifth and sixth, and the ninth and tenth are repetitions of the third and fourth.

68. It has been sometimes stated that the method of the preceding Article is the only one by which Napier's Rules can be demonstrated; this statement, however, is inaccurate, since besides this method Napier himself indicated another method of proof in his *Mirifici Logarithmorum Canonis Descriptio*, pp. 32, 35. This we will now briefly explain.

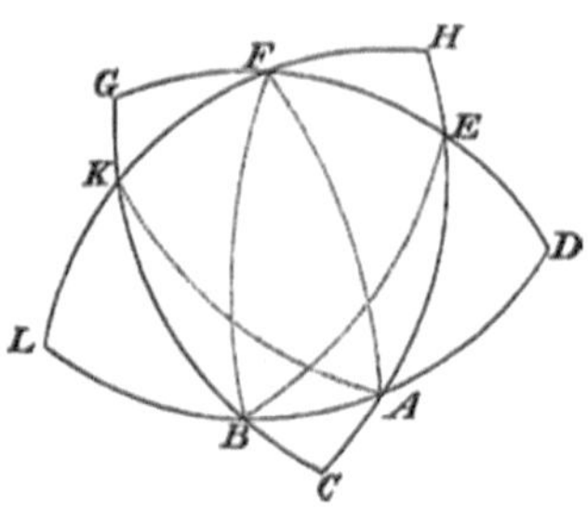

Let ABC be a spherical triangle right-angled at C; with B as pole describe a great circle $DEFG$, and with A as pole describe a great circle $HFKL$, and produce the sides of the original triangle ABC to meet these great circles. Then since B is a pole of $DEFG$ the angles at D and G are right angles, and since A is a pole of $HFKL$ the angles at H and L are right angles. Hence the five triangles BAC, AED, EFH, FKG, KBL are all *right-angled*; and moreover it will be found on examination that, although the elements of these triangles are different, yet *their circular parts are the same.* We will consider, for example, the triangle AED; the angle EAD is equal to the angle BAC, the side AD is the complement of AB; as the angles at C and G are right angles E is a pole of GC (Art. 13), therefore EA is the complement of AC; as B is a pole of DE the angle BED is a right angle, therefore the angle AED is the complement of the angle BEC, that is, the angle AED is the complement of the side BC (Art. 12); and similarly the side DE is equal to the angle DBE, and is therefore the complement of the angle ABC. Hence, if we denote the elements of the triangle ABC as usual by a, b, c, A, B, we have in the triangle AED the hypotenuse equal to $\frac{\pi}{2} - b$, the angles equal to A and $\frac{\pi}{2} - a$, and the sides respectively opposite these angles equal to $\frac{\pi}{2} - B$ and $\frac{\pi}{2} - c$. The *circular parts* of AED are therefore the same as those of ABC. Similarly the remaining three of the five right-angled triangles may be shewn to have the same circular parts as the triangle ABC has.

Now take *two* of the theorems in Art. 65, for example (1) and (3); then the truth of the *ten* cases comprised in Napier's Rules will be found to follow from applying the two theorems in succession to the five triangles formed in the preceding figure. Thus this method of considering Napier's Rules regards each Rule, not as the statement of dissimilar properties of one triangle, but as the statement of similar properties of five allied triangles.

69. In Napier's work a figure is given of which that in the preceding Article is a copy, except that different letters are used; Napier briefly intimates that the truth of the Rules can be easily seen by means of this figure, as well as by the method of induction from consideration of all the cases which can occur. The late T. S. Davies, in his edition of Dr Hutton's *Course of Mathematics*, drew attention to Napier's own views and expanded the demonstration by a systematic examination of the figure of the preceding Article.

It is however easy to evade the necessity of examining the whole figure; all that is wanted is to observe the connexion between the triangle AED and the triangle BAC. For let a_1, a_2, a_3, a_4, a_5 represent the elements of the triangle BAC taken in order, beginning with the hypotenuse and omitting the right angle; then the elements of the triangle AED taken in order, beginning with the hypotenuse and omitting the right angle, are $\frac{\pi}{2}-a_3$, $\frac{\pi}{2}-a_4$, $\frac{\pi}{2}-a_5$, $\frac{\pi}{2}-a_1$, and a_2. If, therefore, to characterise the former we introduce a new set of quantities p_1, p_2, p_3, p_4, p_5, such that $a_1+p_1=a_2+p_2=a_5+p_5=\frac{\pi}{2}$, and that $p_3=a_3$ and $p_4=a_4$, then the original triangle being characterised by p_1, p_2, p_3, p_4, p_5, the second triangle will be similarly characterised by p_3, p_4, p_5, p_1, p_2. As the second triangle can give rise to a third in like manner, and so on, we see that every right-angled triangle is one of a system of five such triangles which are all characterised by the quantities p_1, p_2, p_3, p_4, p_5, always taken in order, each quantity in its turn standing first.

The late R. L. Ellis pointed out this connexion between the five triangles, and thus gave the true significance of Napier's Rules. The memoir containing Mr Ellis's investigations, which was unpublished when the first edition of the present work appeared, will be found in pages 328...335 of *The Mathematical and other writings of Robert Leslie Ellis*...Cambridge, 1863.

Napier's own method of considering his Rules was neglected by writers on the subject until the late T. S. Davies drew attention to it. Hence, as we have already remarked in Art. 68, an erroneous statement was made respecting the Rules. For instance, Woodhouse says, in his *Trigonometry*: "There is no separate and independent proof of these rules;...." Airy says, in the treatise on Trigonometry in the *Encyclopædia Metropolitana*: "These rules are proved to be true only by showing that they comprehend all the equations which we have just found."

70. Opinions have differed with respect to the *utility* of Napier's Rules in practice. Thus Woodhouse says, "In the whole compass of mathematical science there cannot be found, perhaps, rules which more completely attain that which is the proper object of rules, namely, facility and brevity of computation." (*Trigonometry*, chap. X.) On the other hand may be set the following sentence from Airy's Trigonometry (*Encyclopædia Metropolitana*): "In the opinion of Delambre (and no one was better qualified by experience to give an opinion) these theorems are best recollected by the practical calculator in their unconnected form." See Delambre's *Astronomie*, vol. I. p. 205. Professor De Morgan strongly objects to Napier's Rules, and says (*Spherical Trigonometry*, Art. 17): "There are certain mnemonical formulæ called *Napier's Rules of Circular Parts*, which are generally explained. We do not give them, because we are convinced that they only create confusion instead of assisting the memory."

71. We shall now proceed to apply the formulæ of Art. 62 to the solution of right-angled triangles. We shall assume that the given quantities are subject to the limitations which are stated in Arts. 22 and 23, that is, a given side must be less than the semicircumference of a great circle, and a given angle less than two right angles. There will be six cases to consider.

72. *Having given the hypotenuse* c *and an angle* A.

Here we have from (3), (5) and (2) of Art. 62,

$$\tan b = \tan c \cos A, \quad \cot B = \cos c \tan A, \quad \sin a = \sin c \sin A.$$

Thus b and B are determined immediately without ambiguity; and as a must be of the same affection as A (Art. 64), a also is determined without ambiguity.

It is obvious from the formulæ of solution, that in this case the triangle is always possible.

If c and A are both right angles, a is a right angle, and b and B are indeterminate.

73. *Having given a side* b *and the adjacent angle* A.

Here we have from (3), (4) and (6) of Art. 62,

$$\tan c = \frac{\tan b}{\cos A}, \quad \tan a = \tan A \sin b, \quad \cos B = \cos b \sin A.$$

Here c, a, B are determined without ambiguity, and the triangle is always possible.

74. *Having given the two sides* a *and* b.

Here we have from (1) and (4) of Art. 62,

$$\cos c = \cos a \cos b, \quad \cot A = \cot a \sin b, \quad \cot B = \cot b \sin a.$$

Here c, A, B are determined without ambiguity, and the triangle is always possible.

75. *Having given the hypotenuse* c *and a side* a.

Here we have from (1), (3) and (2) of Art. 62,

$$\cos b = \frac{\cos c}{\cos a}, \quad \cos B = \frac{\tan a}{\tan c}, \quad \sin A = \frac{\sin a}{\sin c}.$$

Here b, B, A are determined without ambiguity, since A must be of the same affection as a. It will be seen from these formulæ that there are limitations of the data in order to insure a possible triangle; in fact, c must lie between a and $\pi - a$ in order that the values found for $\cos b$, $\cos B$, and $\sin A$ may be numerically not greater than unity.

If c and a are right angles, A is a right angle, and b and B are indeterminate.

76. *Having given the two angles* A *and* B.

Here we have from (5) and (6) of Art. 62,

$$\cos c = \cot A \cot B, \quad \cos a = \frac{\cos A}{\sin B}, \quad \cos b = \frac{\cos B}{\sin A}.$$

Here c, a, b are determined without ambiguity. There are limitations of the data in order to insure a possible triangle. First suppose A less than $\frac{\pi}{2}$, then B must lie between $\frac{\pi}{2} - A$ and $\frac{\pi}{2} + A$; next suppose A greater than $\frac{\pi}{2}$, then B must lie between $\frac{\pi}{2} - (\pi - A)$ and $\frac{\pi}{2} + (\pi - A)$, that is, between $A - \frac{\pi}{2}$ and $\frac{3\pi}{2} - A$.

77. *Having given a side* a *and the opposite angle* A.

Here we have from (2), (4) and (6) of Art. 62,

$$\sin c = \frac{\sin a}{\sin A}, \qquad \sin b = \tan a \cot A, \qquad \sin B = \frac{\cos A}{\cos a}.$$

Here there is an ambiguity, as the parts are determined from their sines. If $\sin a$ be less than $\sin A$, there are two values admissible for c; corresponding to each of these there will be *in general* only one admissible value of b, since we must have $\cos c = \cos a \cos b$, and only one admissible value of B, since we must have $\cos c = \cot A \cot B$. Thus if one triangle exists with the given parts, there will be *in general* two, and only two, triangles with the given parts. We say *in general* in the preceding sentences, because if $a = A$ there will be only *one* triangle, unless a and A are each right angles, and then b and B become indeterminate.

It is easy to see from a figure that the ambiguity must occur in general.

For, suppose BAC to be a triangle which satisfies the given conditions; produce AB and AC to meet again at A'; then the triangle $A'BC$ also satisfies the given conditions, for it has a right angle at C, BC the given side, and $A' = A$ the given angle.

If $a = A$, then the formulæ of solution shew that c, b, and B are right angles; in this case A is the pole of BC, and the triangle $A'BC$ is symmetrically equal to the triangle ABC (Art. 57).

If a and A are both right angles, B is the pole of AC; B and b are then equal, but may have any value whatever.

There are limitations of the data in order to insure a possible triangle. A

and a must have the same affection by Art. 64; hence the formulæ of solution shew that a must be less than A if both are acute, and greater than A if both are obtuse.

EXAMPLES.

If ABC be a triangle in which the angle C is a right angle, prove the following relations contained in Examples 1 to 5.

1. $\text{Sin}^2 \frac{c}{2} = \sin^2 \frac{a}{2} \cos^2 \frac{b}{2} + \cos^2 \frac{a}{2} \sin^2 \frac{b}{2}$.

2. $\text{Tan}\, \frac{1}{2}(c+a) \tan \frac{1}{2}(c-a) = \tan^2 \frac{b}{2}$.

3. $\text{Sin}(c-b) = \tan^2 \frac{A}{2} \sin(c+b)$.

4. $\text{Sin}\, a \tan \frac{1}{2}A - \sin b \tan \frac{1}{2}B = \sin(a-b)$.

5.
$$\text{Sin}(c-a) = \sin b \cos a \tan \tfrac{1}{2}B,$$
$$\text{Sin}(c-a) = \tan b \cos c \tan \tfrac{1}{2}B.$$

6. If ABC be a spherical triangle, right-angled at C, and $\cos A = \cos^2 a$, shew that if A be not a right angle $b + c = \frac{1}{2}\pi$ or $\frac{3}{2}\pi$, according as b and c are both less or both greater than $\frac{\pi}{2}$.

7. If α, β be the arcs drawn from the right angle respectively perpendicular to and bisecting the hypotenuse c, shew that

$$\sin^2 \frac{c}{2}(1 + \sin^2 \alpha) = \sin^2 \beta.$$

8. In a triangle, if C be a right angle and D the middle point of AB, shew that

$$4\cos^2 \frac{c}{2} \sin^2 CD = \sin^2 a + \sin^2 b.$$

9. In a right-angled triangle, if δ be the length of the arc drawn from C perpendicular to the hypotenuse AB, shew that

$$\cot \delta = \sqrt{(\cot^2 a + \cot^2 b)}.$$

10. OAA_1 is a spherical triangle right-angled at A_1 and acute-angled at A; the arc A_1A_2 of a great circle is drawn perpendicular to OA, then A_2A_3 is drawn perpendicular to OA_1, and so on: shew that A_nA_{n+1} vanishes when n becomes infinite; and find the value of $\cos AA_1 \cos A_1A_2 \cos A_2A_3 \ldots\ldots$ to infinity.

11. ABC is a right-angled spherical triangle, A not being the right angle: shew that if $A = a$, then c and b are quadrants.

12. If δ be the length of the arc drawn from C perpendicular to AB in *any* triangle, shew that

$$\cos\delta = \operatorname{cosec} c\,(\cos^2 a + \cos^2 b - 2\cos a\,\cos b\,\cos c)^{\frac{1}{2}}.$$

13. ABC is a great circle of a sphere; AA', BB', CC', are arcs of great circles drawn at right angles to ABC and reckoned positive when they lie on the same side of it: shew that the condition of A', B', C' lying in a great circle is

$$\tan AA'\,\sin BC + \tan BB'\,\sin CA + \tan CC'\,\sin AB = 0.$$

14. Perpendiculars are drawn from the angles A, B, C of any triangle meeting the opposite sides at D, E, F respectively: shew that

$$\tan BD\,\tan CE\,\tan AF = \tan DC\,\tan EA\,\tan FB.$$

15. Ox, Oy are two great circles of a sphere at right angles to each other, P is any point in AB another great circle. $OC = p$ is the arc perpendicular to AB from O, making the angle $COx = a$ with Ox. PM, PN are arcs perpendicular to Ox, Oy respectively: shew that if $OM = x$ and $ON = y$,

$$\cos a\,\tan x + \sin a\,\tan y = \tan p.$$

16. The position of a point on a sphere, with reference to two great circles at right angles to each other as axes, is determined by the portions θ, ϕ of these circles cut off by great circles through the point, and through two points on the axes, each $\frac{\pi}{2}$ from their point of intersection: shew that if the three points (θ, ϕ), (θ', ϕ'), (θ'', ϕ'') lie on the same great circle

$$\tan\phi\,(\tan\theta' - \tan\theta'') + \tan\phi'\,(\tan\theta'' - \tan\theta)$$

$$+\tan\phi''(\tan\theta - \tan\theta') = 0.$$

17. If a point on a sphere be referred to two great circles at right angles to each other as axes, by means of the portions of these axes cut off by great circles drawn through the point and two points on the axes each 90° from their intersection, shew that the equation to a great circle is

$$\tan\theta\cot\alpha + \tan\phi\cot\beta = 1.$$

18. In a spherical triangle, if $A = \frac{\pi}{5}$, $B = \frac{\pi}{3}$, and, $C = \frac{\pi}{2}$, shew that $a + b + c = \frac{\pi}{2}$.

VI

SOLUTION OF OBLIQUE-ANGLED TRIANGLES.

78. The solution of oblique-angled triangles may be made in some cases to depend immediately on the solution of right-angled triangles; we will indicate these cases before considering the subject generally.

(1) Suppose a triangle to have one of its given sides equal to a *quadrant*. In this case the polar triangle has its corresponding angle a right angle; the polar triangle can therefore be solved by the rules of the preceding Chapter, and thus the elements of the primitive triangle become known.

(2) Suppose among the given elements of a triangle there are two *equal sides* or two *equal angles*. By drawing an arc from the vertex to the middle point of the base, the triangle is divided into two equal *right-angled* triangles; by the solution of one of these right-angled triangles the required elements can be found.

(3) Suppose among the given elements of a triangle there are two sides, one of which is the supplement of the other, or two angles, one of which is the supplement of the other. Suppose, for example, that $b + c = \pi$, or else that $B + C = \pi$; produce BA and BC to meet at B' (see the first figure to Art. 38); then the triangle $B'AC$ has two equal sides given, or else two equal angles

given; and by the preceding case the solution of it can be made to depend on the solution of a right-angled triangle.

79. We now proceed to the solution of oblique-angled triangles in general. There will be six cases to consider.

80. *Having given the three sides.*

Here we have $\cos A = \dfrac{\cos a - \cos b \cos c}{\sin b \sin c}$, and similar formulæ for $\cos B$ and $\cos C$. Or if we wish to use formulæ suited to logarithms, we may take the formula for the sine, cosine, or tangent of half an angle given in Art. 45. In selecting a formula, attention should be paid to the remarks in *Plane Trigonometry*, Chap. XII. towards the end.

81. *Having given the three angles.*

Here we have $\cos a = \dfrac{\cos A + \cos B \cos C}{\sin B \sin C}$, and similar formulæ for $\cos b$ and $\cos c$. Or if we wish to use formulæ suited to logarithms, we may take the formula for the sine, cosine, or tangent of half a side given in Art. 49.

There is no ambiguity in the two preceding cases; the triangles however may be impossible with the given elements.

82. *Having given two sides and the included angle* (a, C, b).

By Napier's analogies

$$\tan \tfrac{1}{2}(A + B) = \frac{\cos \frac{1}{2}(a - b)}{\cos \frac{1}{2}(a + b)} \cot \tfrac{1}{2}C,$$

$$\tan \tfrac{1}{2}(A - B) = \frac{\sin \frac{1}{2}(a - b)}{\sin \frac{1}{2}(a + b)} \cot \tfrac{1}{2}C;$$

these determine $\frac{1}{2}(A + B)$ and $\frac{1}{2}(A - B)$, and thence A and B.

Then c may be found from the formula $\sin c = \dfrac{\sin a \sin C}{\sin A}$; in this case, since c is found from its sine, it may be uncertain which of two values is to be given to it; the point may be sometimes settled by observing that the greater side of a triangle is opposite to the greater angle. Or we may determine c from equation (1) of Art. 54, which is free from ambiguity.

Or we may determine c, without previously determining A and B, from the formula $\cos c = \cos a \cos b + \sin a \sin b \cos C$; this is free from ambiguity. This formula may be adapted to logarithms thus:

$$\cos c = \cos b\,(\cos a + \sin a \tan b \cos C);$$

assume $\tan\theta = \tan b\, \cos C$; then

$$\cos c = \cos b\,(\cos a + \sin a\, \tan\theta) = \frac{\cos b\, \cos(a-\theta)}{\cos\theta};$$

this is adapted to logarithms.

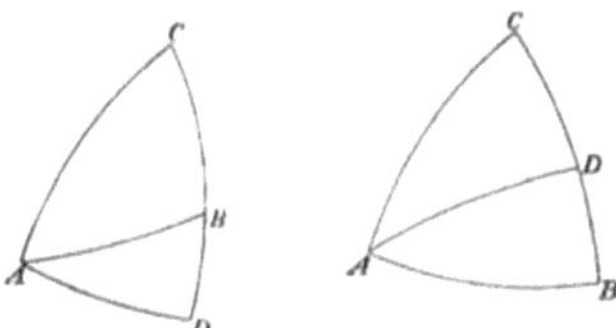

Or we may treat this case conveniently by resolving the triangle into the sum or difference of two right-angled triangles. From A draw the arc AD perpendicular to CB or CB produced; then, by Art. 62, $\tan CD = \tan b \cos C$, and this determines CD, and then DB is known. Again, by Art. 62,

$$\cos c = \cos AD \cos DB = \cos DB \frac{\cos b}{\cos CD};$$

this finds c. It is obvious that CD is what was denoted by θ in the former part of the Article.

By Art. 62,

$$\tan AD = \tan C \sin CD, \text{ and } \tan AD = \tan ABD \sin DB;$$
$$\text{thus } \tan ABD \sin DB = \tan C \sin\theta,$$

where $DB = a - \theta$ or $\theta - a$, according as D is on CB or CB produced, and ABD is either B or the supplement of B; this formula enables us to find B independently of A.

Thus, in the present case, there is no real ambiguity, and the triangle is always possible.

83. *Having given two angles and the included side* (A, c, B).

By Napier's analogies,

$$\tan\tfrac{1}{2}(a+b) = \frac{\cos\frac{1}{2}(A-B)}{\cos\frac{1}{2}(A+B)}\tan\tfrac{1}{2}c,$$

$$\tan\tfrac{1}{2}(a-b) = \frac{\sin\frac{1}{2}(A-B)}{\sin\frac{1}{2}(A+B)}\tan\tfrac{1}{2}c;$$

these determine $\frac{1}{2}(a+b)$ and $\frac{1}{2}(a-b)$, and thence a and b.

Then C may be found from the formula $\sin C = \dfrac{\sin A \sin c}{\sin a}$; in this case, since C is found from its sine, it may be uncertain which of two values is to be given to it; the point may be sometimes settled by observing that the greater angle of a triangle is opposite to the greater side. Or we may determine C from equation (3) of Art. 54, which is free from ambiguity.

Or we may determine C without previously determining a and b from the formula $\cos C = -\cos A \cos B + \sin A \sin B \cos c$. This formula may be adapted to logarithms, thus:

$$\cos C = \cos B(-\cos A + \sin A \tan B \cos c);$$

assume $\cot\phi = \tan B \cos c$; then

$$\cos C = \cos B(-\cos A + \cot\phi \sin A) = \frac{\cos B \sin(A-\phi)}{\sin\phi};$$

this is adapted to logarithms.

Or we may treat this case conveniently by resolving the triangle into the sum or difference of two right-angled triangles. From A draw the arc AD perpendicular to CB (see the right-hand figure of Art. 82); then, by Art. 62, $\cos c = \cot B \cot DAB$, and this determines DAB, and then CAD is known. Again, by Art. 62,

$$\cos AD \sin CAD = \cos C, \text{ and } \cos AD \sin BAD = \cos B;$$

therefore $\dfrac{\cos C}{\sin CAD} = \dfrac{\cos B}{\sin BAD}$; this finds C.
It is obvious that DAB is what was denoted by ϕ in the former part of the Article.

By Art. 62,

$$\tan AD = \tan AC \cos CAD, \text{ and } \tan AD = \tan AB \cos BAD;$$

thus
$$\tan b \cos CAD = \tan c \cos \phi,$$

where $CAD = A - \phi$; this formula enables us to find b independently of a.

Similarly we may proceed when the perpendicular AD falls on CB *produced*; (see the left-hand figure of Art. 82).

Thus, in the present case, there is no real ambiguity; moreover the triangle is always possible.

84. *Having given two sides and the angle opposite one of them* (a, b, A).

The angle B may be found from the formula

$$\sin B = \frac{\sin b}{\sin a} \sin A;$$

and then C and c may be found from Napier's analogies,

$$\tan \tfrac{1}{2}C = \frac{\cos \frac{1}{2}(a-b)}{\cos \frac{1}{2}(a+b)} \cot \tfrac{1}{2}(A+B),$$

$$\tan \tfrac{1}{2}c = \frac{\cos \frac{1}{2}(A+B)}{\cos \frac{1}{2}(A-B)} \tan \tfrac{1}{2}(a+b).$$

In this case, since B is found from its sine, there will sometimes be two solutions; and sometimes there will be no solution at all, namely, when the value found for $\sin B$ is greater than unity. We will presently return to this point. (See Art. 86.)

We may also determine C and c independently of B by formulæ adapted to logarithms. For, by Art. 44,

$$\cot a \sin b = \cos b \cos C + \sin C \cot A = \cos b \left(\cos C + \frac{\cot A}{\cos b} \sin C\right);$$

assume $\tan \phi = \dfrac{\cot A}{\cos b}$; thus

$$\cot a \sin b = \cos b(\cos C + \tan \phi \sin C) = \frac{\cos b \cos(C - \phi)}{\cos \phi};$$

therefore
$$\cos(C - \phi) = \cos \phi \cot a \tan b;$$

from this equation $C - \phi$ is to be found, and then C. The ambiguity still exists; for if the last equation leads to $C - \phi = \alpha$, it will be satisfied also by $\phi - C = \alpha$; so that we have two admissible values for C, if $\phi + \alpha$ is less than π, and $\phi - \alpha$ is positive.

And

$$\cos a = \cos b \cos c + \sin b \sin c \cos A = \cos b(\cos c + \sin c \tan b \cos A);$$

assume $\tan\theta = \tan b \cos A$; thus

$$\cos a = \cos b(\cos c + \sin c \tan\theta) = \frac{\cos b \cos(c - \theta)}{\cos\theta};$$

therefore $$\cos(c - \theta) = \frac{\cos a \cos\theta}{\cos b};$$

from this equation $c - \theta$ is to be found, and then c; and there may be an ambiguity as before.

Or we may treat this case conveniently by resolving the triangle into the sum or difference of two right-angled triangles.

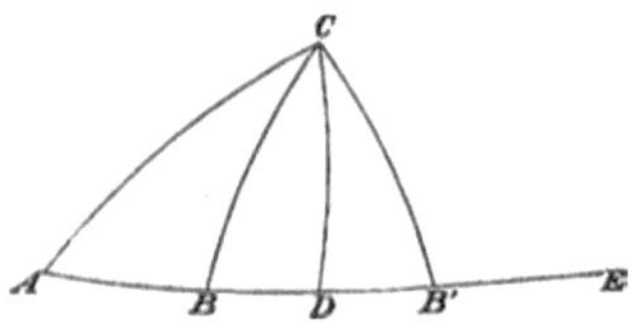

Let $CA = b$, and let $CAE =$ the given angle A; from C draw CD perpendicular to AE, and let CB and $CB' = a$; thus the figure shews that there may be two triangles which have the given elements. Then, by Art. 62, $\cos b = \cot A \cot ACD$; this finds ACD. Again, by Art. 62,

$$\tan CD = \tan AC \cos ACD,$$
$$\text{and } \tan CD = \tan CB \cos BCD, \text{ or } \tan CB' \cos B'CD,$$

therefore $\tan AC \cos ACD = \tan CB \cos BCD$, or $\tan CB' \cos B'CD$; this finds BCD or $B'CD$.

It is obvious that ACD is what was denoted by ϕ in the former part of the Article.

Also, by Art. 62, $\tan AD = \tan AC \cos A$; this finds AD. Then

$$\cos AC = \cos CD \cos AD,$$

$$\cos CB = \cos CD \cos BD,$$

$$\text{or } \cos CB' = \cos CD \cos B'D;$$

therefore
$$\frac{\cos AC}{\cos AD} = \frac{\cos CB}{\cos BD} \text{ or } \frac{\cos CB'}{\cos B'D};$$

this finds BD or $B'D$.

It is obvious that AD is what was denoted by θ in the former part of the Article.

85. *Having given two angles and the side opposite one of them* (A, B, a).

This case is analogous to that immediately preceding, and gives rise to the same ambiguities. The side b may be found from the formula

$$\sin b = \frac{\sin B \sin a}{\sin A};$$

and then C and c may be found from Napier's analogies,

$$\tan \tfrac{1}{2}C = \frac{\cos \frac{1}{2}(a-b)}{\cos \frac{1}{2}(a+b)} \cot \tfrac{1}{2}(A+B),$$

$$\tan \tfrac{1}{2}c = \frac{\cos \frac{1}{2}(A+B)}{\cos \frac{1}{2}(A-B)} \tan \tfrac{1}{2}(a+b),$$

We may also determine C and c independently of b by formulæ adapted to logarithms. For

$$\cos A = -\cos B \cos C + \sin B \sin C \cos a$$

$$= \cos B(-\cos C + \tan B \sin C \cos a),$$

assume $\cot \phi = \tan B \cos a$; thus

$$\cos A = \cos B(-\cos C + \sin C \cot \phi) = \frac{\cos B \sin(C-\phi)}{\sin \phi};$$

therefore $$\sin(C - \phi) = \frac{\cos A \sin \phi}{\cos B};$$

from this equation $C - \phi$ is to be found and then C. Since $C - \phi$ is found from its sine there may be an ambiguity. Again, by Art. 44,

$$\cot A \sin B = \cot a \sin c - \cos c \cos B = \cos B \left(- \cos c + \frac{\cot a \sin c}{\cos B} \right),$$

assume $\cot \theta = \dfrac{\cot a}{\cos B}$; then

$$\cot A \sin B = \cos B(-\cos c + \sin c \cot \theta) = \frac{\cos B \sin(c - \theta)}{\sin \theta};$$

therefore $$\sin(c - \theta) = \cot A \tan B \sin \theta;$$

from this equation $c - \theta$ is to be found, and then c. Since $c - \theta$ is found from its sine there may be an ambiguity. As before, it may be shewn that these results agree with those obtained by resolving the triangle into two right-angled triangles; for if in the triangle ACB' the arc CD be drawn perpendicular to AB', then $B'CD$ will $= \phi$, and $B'D = \theta$.

86. We now return to the consideration of the ambiguity which may occur in the case of Art. 84, when two sides are given and the angle opposite one of them. The discussion is somewhat tedious from its length, but presents no difficulty.

Before considering the problem generally, we will take the particular case in which $a = b$; then A must $= B$. The first and third of Napier's analogies give

$$\cot \tfrac{1}{2}C = \tan A \cos a, \qquad \tan \tfrac{1}{2}c = \tan a \cos A;$$

now $\cot \frac{1}{2}C$ and $\tan \frac{1}{2}c$ must both be *positive*, so that A and a must be of the same affection. Hence, when $a = b$, there will be no solution at all, unless A and a are of the same affection, and then there will be only one solution; except when A and a are both right angles, and then $\cot \frac{1}{2}C$ and $\tan \frac{1}{2}c$ are indeterminate, and there is an infinite number of solutions.

We now proceed to the general discussion.

If $\sin b \sin A$ be greater than $\sin a$, there is no triangle which satisfies the given conditions; if $\sin b \sin A$ is *not* greater than $\sin a$, the equation $\sin B = \dfrac{\sin b \sin A}{\sin a}$ furnishes two values of B, which we will denote by β and β', so that $\beta' = \pi - \beta$;

we will suppose that β is the one which is not greater than the other.

Now, in order that these values of B may be admissible, it is necessary and sufficient that the values of $\cot \frac{1}{2}C$ and of $\tan \frac{1}{2}c$ should both be positive, that is, $A-B$ and $a-b$ must have the same sign by the second and fourth of Napier's analogies. We have therefore to compare the sign of $A-\beta$ and the sign of $A-\beta'$ with that of $a-b$.

We will suppose that A is less than a right angle, and separate the corresponding discussion into three cases.

I. Let b be less than $\dfrac{\pi}{2}$.

(1) Let a be less than b; the formula $\sin B = \dfrac{\sin b}{\sin a} \sin A$ make β greater than A, and *à fortiori* β' greater than A. Hence there are two solutions.

(2) Let a be equal to b; then there is one solution, as previously shewn.

(3) Let a be greater than b; we may have then $a+b$ less than π or equal to π or greater than π. If $a+b$ is less than π, then $\sin a$ is greater than $\sin b$; thus β is less than A and therefore admissible, and β' is greater than A and inadmissible. Hence there is one solution. If $a+b$ is equal to π, then β is equal to A, and β' greater than A, and both are inadmissible. Hence there is no solution. If $a+b$ is greater than π, then $\sin a$ is less than $\sin b$, and β and β' are both greater than A, and both inadmissible. Hence there is no solution.

II. Let b be equal to $\dfrac{\pi}{2}$.

(1) Let a be less than b; then β and β' are both greater than A, and both admissible. Hence there are two solutions.

(2) Let a be equal to b; then there is no solution, as previously shewn.

(3) Let a be greater than b; then $\sin a$ is less than $\sin b$, and β and β' are both greater than A, and inadmissible. Hence there is no solution.

III. Let b be greater than $\dfrac{\pi}{2}$.

(1) Let a be less than b; we may have then $a+b$ less than π or equal to π or greater than π. If $a+b$ is less than π, then $\sin a$ is less than $\sin b$, and β and β' are both greater than A and both admissible. Hence there are two solutions. If $a+b$ is equal to π, then β is equal to A and inadmissible, and β' is greater than A and admissible. Hence there is one solution. If $a+b$ is greater than π, then $\sin a$ is greater than $\sin b$; β is less than A and admissible, and β' is greater than A and admissible. Hence there is one solution.

(2) Let a be equal to b; then there is no solution, as previously shewn.

(3) Let a be greater than b; then $\sin a$ is less than $\sin b$, and β and β' are both greater than A and both inadmissible. Hence there is no solution.

We have then the following results when A *is less than a right angle.*

$$b < \frac{\pi}{2}\begin{cases} a < b \ldots\ldots\ldots\ldots\ldots\ldots\ldots\ldots \text{two solutions,} \\ a = b \ldots\ldots\ldots\ldots\ldots\ldots\ldots\ldots \text{one solution,} \\ a > b \text{ and } a + b < \pi \ldots\ldots\ldots\ldots \text{one solution,} \\ a > b \text{ and } a + b = \pi \text{ or } > \pi \ldots\ldots \text{no solution.} \end{cases}$$

$$b = \frac{\pi}{2}\begin{cases} a < b \ldots\ldots\ldots\ldots\ldots\ldots\ldots\ldots \text{two solutions,} \\ a = b \text{ or } a > b \ldots\ldots\ldots\ldots\ldots \text{no solution.} \end{cases}$$

$$b > \frac{\pi}{2}\begin{cases} a < b \text{ and } a + b < \pi \ldots\ldots\ldots\ldots \text{two solutions,} \\ a < b \text{ and } a + b = \pi \text{ or } > \pi \ldots\ldots \text{one solution,} \\ a = b \text{ or } > b \ldots\ldots\ldots\ldots\ldots\ldots \text{no solution.} \end{cases}$$

It must be remembered, however, that in the cases in which two solutions are indicated, there will be no solution at all if $\sin a$ be less than $\sin b \sin A$.

In the same manner the cases in which A is equal to a right angle or greater than a right angle may be discussed, and the following results obtained.

When A *is equal to a right angle,*

$$b < \frac{\pi}{2}\begin{cases} a < b \text{ or } a = b \ldots\ldots\ldots\ldots\ldots\ldots \text{no solution,} \\ a > b \text{ and } a + b < \pi \ldots\ldots\ldots\ldots \text{one solution,} \\ a > b \text{ and } a + b = \pi \text{ or } > \pi \ldots\ldots \text{no solution.} \end{cases}$$

$$b = \frac{\pi}{2}\begin{cases} a < b \text{ or } a > b \ldots\ldots\ldots\ldots\ldots\ldots \text{no solution,} \\ a = b \ldots\ldots\ldots\ldots \text{infinite number of solutions.} \end{cases}$$

$$b > \frac{\pi}{2}\begin{cases} a < b \text{ and } a + b > \pi \ldots\ldots\ldots\ldots \text{one solution,} \\ a < b \text{ and } a + b = \pi \text{ or } < \pi \ldots\ldots \text{no solution,} \\ a = b \text{ or } a > b \ldots\ldots\ldots\ldots\ldots\ldots \text{no solution.} \end{cases}$$

When A *is greater than a right angle,*

$$b < \frac{\pi}{2} \begin{cases} a < b \text{ or } a = b \text{ .... no solution,} \\ a > b \text{ and } a + b = \pi \text{ or } < \pi \text{ .... one solution,} \\ a > b \text{ and } a + b > \pi \text{ .... two solutions.} \end{cases}$$

$$b = \frac{\pi}{2} \begin{cases} a < b \text{ or } a = b \text{ .... no solution,} \\ a > b \text{ .... two solutions.} \end{cases}$$

$$b > \frac{\pi}{2} \begin{cases} a < b \text{ and } a + b > \pi \text{ .... one solution,} \\ a < b \text{ and } a + b = \pi \text{ or } < \pi \text{ .... no solution,} \\ a = b \text{ .... one solution,} \\ a > b \text{ .... two solutions.} \end{cases}$$

As before in the cases in which two solutions are indicated, there will be no solution at all if $\sin a$ be less than $\sin b \sin A$.

It will be seen from the above investigations that if a lies between b and $\pi - b$, there will be one solution; if a does not lie between b and $\pi - b$ either there are two solutions or there is no solution; this enunciation is not meant to include the cases in which $a = b$ or $= \pi - b$.

87. The results of the preceding Article may be illustrated by a figure.

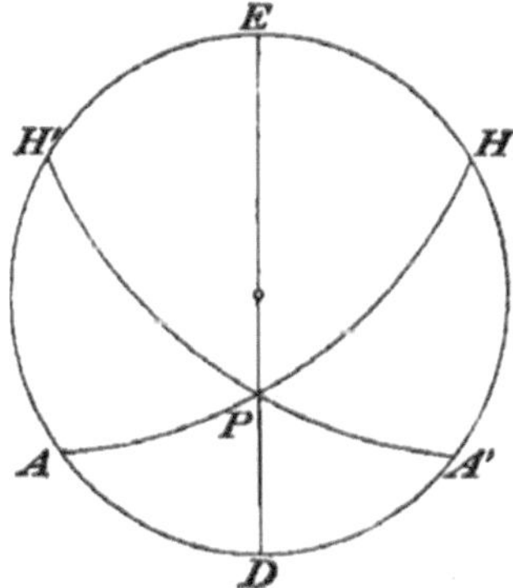

Let $ADA'E$ be a great circle; suppose PA and PA' the projections on the plane of this circle of arcs which are each equal to b and inclined at an angle A to ADA'; let PD and PE be the projections of the least and greatest distances of P from the great circle (see Art. 59). Thus the figure supposes A and b each less than $\frac{\pi}{2}$.

If a be less than the arc which is represented by PD there is no triangle; if a be between PD and PA in magnitude, there are two triangles, since B will fall on ADA', and we have two triangles BPA and BPA'; if a be between PA and PH there will be only one triangle, as B will fall on $A'H$ or AH', and the triangle will be either APB with B between A' and H, or else $A'PB$ with B between A and H'; but these two triangles are symmetrically equal (Art. 57); if a be greater than PH there will be no triangle. The figure will easily serve for all the cases; thus if A is greater than $\frac{\pi}{2}$, we can suppose PAE and $PA'E$ to be equal to A; if b is greater than $\frac{\pi}{2}$, we can take PH and PH' to represent b.

88. The ambiguities which occur in the last case in the solution of oblique-angled triangles (Art. 85) may be discussed in the same manner as those in Art. 86; or, by means of the polar triangle, the last case may be deduced from that of Art. 86.

EXAMPLES.

1. The sides of a triangle are 105°, 90°, and 75° respectively: find the sines of all the angles.

2. Shew that $\tan\frac{1}{2}A\tan\frac{1}{2}B = \dfrac{\sin(s-c)}{\sin s}$. Solve a triangle when a side, an adjacent angle, and the sum of the other two sides are given.

3. Solve a triangle having given a side, an adjacent angle, and the sum of the other two angles.

4. A triangle has the sum of two sides equal to a semicircumference: find the arc joining the vertex with the middle of the base.

5. If a, b, c are known, c being a *quadrant*, determine the angles: shew also that if δ be the perpendicular on c from the opposite angle, $\cos^2\delta = \cos^2 a + \cos^2 b$.

6. If one side of a spherical triangle be divided into four equal parts, and θ_1, θ_2, θ_3, θ_4, be the angles subtended at the opposite angle by the parts taken in order, shew that

$$\sin(\theta_1+\theta_2)\sin\theta_2\sin\theta_4 = \sin(\theta_3+\theta_4)\sin\theta_1\sin\theta_3.$$

7. In a spherical triangle if $A = B = 2C$, shew that

$$8 \sin\left(a + \frac{c}{2}\right) \sin^2 \frac{c}{2} \cos \frac{c}{2} = \sin^3 a.$$

8. In a spherical triangle if $A = B = 2C$, shew that

$$8 \sin^2 \frac{C}{2} \left(\cos s + \sin \frac{C}{2}\right) \frac{\cos \frac{c}{2}}{\cos a} = 1.$$

9. If the equal sides of an isosceles triangle ABC be bisected by an arc DE, and BC be the base, shew that

$$\sin \frac{DE}{2} = \tfrac{1}{2} \sin \frac{BC}{2} \sec \frac{AC}{2}.$$

10. If c_1, c_2 be the two values of the third side when A, a, b are given and the triangle is ambiguous, shew that

$$\tan \frac{c_1}{2} \tan \frac{c_2}{2} = \tan \tfrac{1}{2}(b - a) \tan \tfrac{1}{2}(b + a).$$

VII

CIRCUMSCRIBED AND INSCRIBED CIRCLES.

89. *To find the angular radius of the small circle inscribed in a given triangle.*

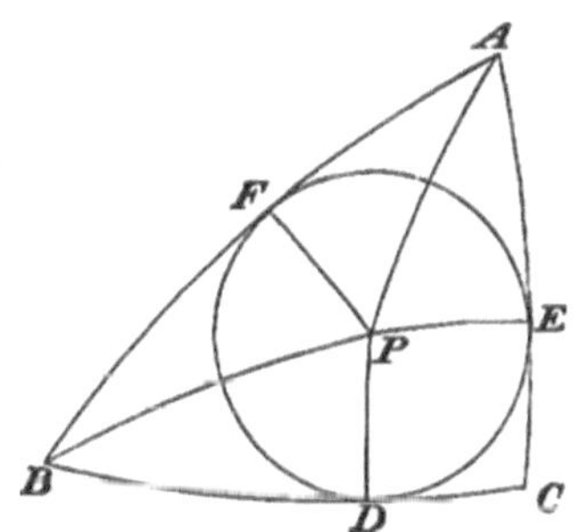

Let ABC be the triangle; bisect the angles A and B by arcs meeting at P; from P draw PD, PE, PF perpendicular to the sides. Then it may be shewn that PD, PE, PF are all equal; also that $AE = AF$, $BF = BD$, $CD = CE$. Hence $BC + AF =$ half the sum of the sides $= s$; therefore $AF = s - a$. Let $PF = r$.

Now $$\tan PF = \tan PAF \sin AF \text{ (Art. 62);}$$

thus $$\tan r = \tan \frac{A}{2} \sin(s - a). \tag{1}$$

The value of $\tan r$ may be expressed in various forms; thus from Art. 45, we obtain

$$\tan\frac{A}{2} = \sqrt{\frac{\sin(s-b)\sin(s-c)}{\sin s\,\sin(s-a)}}\,;$$

substitute this value in (1), thus

$$\tan r = \sqrt{\left\{\frac{\sin(s-a)\sin(s-b)\sin(s-c)}{\sin s}\right\}} = \frac{n}{\sin s}\ \text{(Art. 46)}. \qquad (2)$$

Again

$$\begin{aligned}\sin(s-a) &= \sin\{\tfrac{1}{2}(b+c)-\tfrac{1}{2}a\}\\ &= \sin\tfrac{1}{2}(b+c)\cos\tfrac{1}{2}a - \cos\tfrac{1}{2}(b+c)\sin\tfrac{1}{2}a\\ &= \frac{\sin\frac{1}{2}a\cos\frac{1}{2}a}{\sin\frac{1}{2}A}\{\cos\tfrac{1}{2}(B-C)-\cos\tfrac{1}{2}(B+C)\},\ \text{(Art. 54)}\\ &= \frac{\sin a\sin\frac{1}{2}B\sin\frac{1}{2}C}{\sin\frac{1}{2}A};\end{aligned}$$

therefore from (1) $\tan r = \dfrac{\sin\frac{1}{2}B\sin\frac{1}{2}C}{\cos\frac{1}{2}A}\sin a;$ (3)

hence, by Art. 51,

$$\tan r = \frac{\sqrt{\{-\cos S\cos(S-A)\cos(S-B)\cos(S-C)\}}}{2\cos\frac{1}{2}A\cos\frac{1}{2}B\cos\frac{1}{2}C}$$

$$= \frac{N}{2\cos\frac{1}{2}A\cos\frac{1}{2}B\cos\frac{1}{2}C}. \qquad (4)$$

It may be shewn by common trigonometrical formulæ that

$$4\cos\tfrac{1}{2}A\cos\tfrac{1}{2}B\cos\tfrac{1}{2}C = \cos S+\cos(S-A)+\cos(S-B)+\cos(S-C);$$

hence we have from (4)

$$\cot r = \frac{1}{2N}\{\cos S+\cos(S-A)+\cos(S-B)+\cos(S-C)\}. \qquad (5)$$

90. *To find the angular radius of the small circle described so as to touch one side of a given triangle, and the other sides produced.*

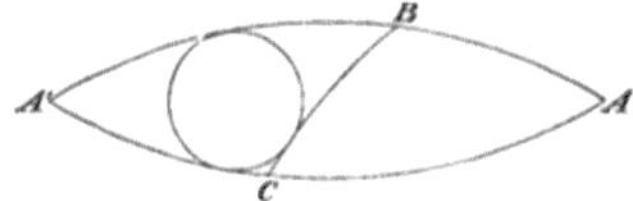

Let ABC be the triangle; and suppose we require the radius of the small circle which touches BC, and AB and AC produced. Produce AB and AC to meet at A'; then we require the radius of the small *circle inscribed in* A'BC, and the sides of $A'BC$ are a, $\pi - b$, $\pi - c$ respectively. Hence if r_1 be the required radius, and s denote as usual $\frac{1}{2}(a+b+c)$, we have from Art. 89,

$$\tan r_1 = \tan\frac{A}{2}\sin s. \tag{1}$$

From this result we may derive other equivalent forms as in the preceding Article; or we may make use of those forms immediately, observing that the angles of the triangle $A'BC$ are A, $\pi - B$, $\pi - C$ respectively. Hence s being $\frac{1}{2}(a+b+c)$ and S being $\frac{1}{2}(A+B+C)$ we shall obtain

$$\tan r_1 = \sqrt{\left\{\frac{\sin s\sin(s-b)\sin(s-c)}{\sin(s-a)}\right\}} = \frac{n}{\sin(s-a)}, \tag{2}$$

$$\tan r_1 = \frac{\cos\frac{1}{2}B\cos\frac{1}{2}C}{\cos\frac{1}{2}A}\sin a, \tag{3}$$

$$\tan r_1 = \frac{\sqrt{\{-\cos S\cos(S-A)\cos(S-B)\cos(S-C)\}}}{2\cos\frac{1}{2}A\sin\frac{1}{2}B\sin\frac{1}{2}C} \tag{4}$$

$$= \frac{N}{2\cos\frac{1}{2}A\sin\frac{1}{2}B\sin\frac{1}{2}C},$$

$$\cot r_1 = \frac{1}{2N}\{-c\cos S - \cos(S-A) + \cos(S-B) + \cos(S-C)\}. \tag{5}$$

These results may also be found independently by bisecting two of the angles of the triangle $A'BC$, so as to determine the pole of the small circle, and proceeding as in Art. 89.

91. A circle which touches one side of a triangle and the other sides produced is called an *escribed circle;* thus there are three escribed circles belonging to a given triangle. We may denote the radii of the escribed circles which touch CA and AB respectively by r_2 and r_3, and values of $\tan r_2$ and $\tan r_3$ may be found from what has been already given with respect to $\tan r_1$ by appropriate changes in the letters which denote the sides and angles.

In the preceding Article a triangle $A'BC$ was formed by producing AB and AC to meet again at A'; similarly another triangle may be formed by producing BC and BA to meet again, and another by producing CA and CB to meet again. The original triangle ABC and the three formed from it have been called *associated triangles*, ABC being the fundamental triangle. Thus the inscribed and escribed circles of a given triangle are the same as the circles inscribed in the system of associated triangles of which the given triangle is the fundamental triangle.

92. *To find the angular radius of the small circle described about a given triangle.*

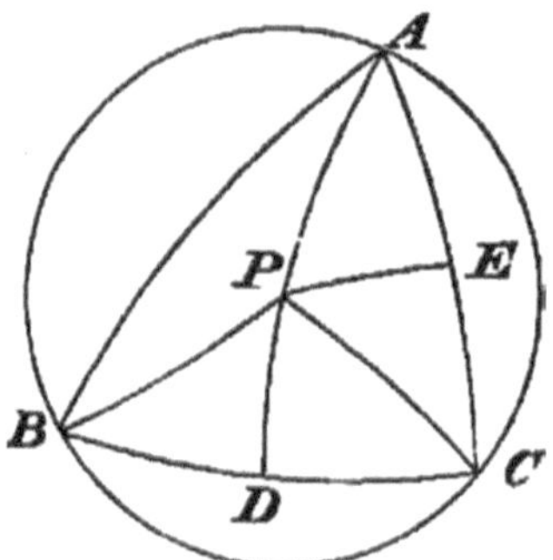

Let ABC be the given triangle; bisect the sides CB, CA at D and E respectively, and draw from D and E arcs at right angles to CB and CA respectively, and let P be the intersection of these arcs. Then P will be the pole of the small circle described about ABC. For draw PA, PB, PC; then from the right-angled triangles PCD and PBD it follows that $PB = PC$; and from the right-angled triangles PCE and PAE it follows that $PA = PC$; hence $PA = PB = PC$. Also the angle $PAB =$ the angle PBA, the angle $PBC =$ the angle PCB, and the angle $PCA =$ the angle PAC; therefore $PCB + A = \frac{1}{2}(A + B + C)$, and $PCB = S - A$.

Let $PC = R$.

Now $$\tan CD = \tan CP \cos PCD, \text{ (Art. 62,)}$$

thus $$\tan \tfrac{1}{2}a = \tan R \cos(S - A),$$

therefore $$\tan R = \frac{\tan \frac{1}{2}a}{\cos(S - A)}. \qquad (1)$$

The value of $\tan R$ may be expressed in various forms; thus if we substitute for $\tan \dfrac{a}{2}$ from Art. 49, we obtain

$$\tan R = \sqrt{\left\{\frac{-\cos S}{\cos(S - A)\cos(S - B)\cos(S - C)}\right\}} = \frac{\cos S}{N}. \qquad (2)$$

Again $$\cos(S - A) = \cos\left\{\tfrac{1}{2}(B + C) - \tfrac{1}{2}A\right\}$$

$$= \cos \tfrac{1}{2}(B + C)\cos \tfrac{1}{2}A + \sin \tfrac{1}{2}(B + C)\sin \tfrac{1}{2}A$$

$$= \frac{\sin \frac{1}{2}A \cos \frac{1}{2}A}{\cos \frac{1}{2}a}\left\{\cos \tfrac{1}{2}(b + c) + \cos \tfrac{1}{2}(b - c)\right\}, \text{ (Art. 54,)}$$

$$= \frac{\sin A}{\cos \frac{1}{2}a}\cos \tfrac{1}{2}b \cos \tfrac{1}{2}c;$$

therefore from (1)

$$\tan R = \frac{\sin \frac{1}{2}a}{\sin A \cos \frac{1}{2}b \cos \frac{1}{2}c}. \qquad (3)$$

Substitute in the last expression the value of $\sin A$ from Art. 46; thus

$$\tan R = \frac{2\sin \frac{1}{2}a \sin \frac{1}{2}b \sin \frac{1}{2}c}{\sqrt{\{\sin s \sin(s - a)\sin(s - b)\sin(s - c)\}}}$$

$$= \frac{2\sin \frac{1}{2}a \sin \frac{1}{2}b \sin \frac{1}{2}c}{n}. \qquad (4)$$

It may be shewn, by common trigonometrical formulæ that

$$4\sin \tfrac{1}{2}a \sin \tfrac{1}{2}b \sin \tfrac{1}{2}c = \sin(s - a) + \sin(s - b) + \sin(s - c) - \sin s;$$

hence we have from (4)

$$\tan R = \frac{1}{2n}\{\sin(s-a) + \sin(s-b) + \sin(s-c) - \sin s\}. \tag{5}$$

93. *To find the angular radii of the small circles described round the triangles associated with a given fundamental triangle.*

Let R_1 denote the radius of the circle described round the triangle formed by producing AB and AC to meet again at A'; similarly let R_2 and R_3 denote the radii of the circles described round the other two triangles which are similarly formed. Then we may deduce expressions for $\tan R_1$, $\tan R_2$, and $\tan R_3$ from those found in Art. 92 for $\tan R$. The sides of the triangle $A'BC$ are a, $\pi - b$, $\pi - c$, and its angles are A, $\pi - B$, $\pi - C$; hence if $s = \frac{1}{2}(a+b+c)$ and $S = \frac{1}{2}(A+B+C)$ we shall obtain from Art. 92

$$\tan R_1 = \frac{\tan\frac{1}{2}a}{-\cos S}, \tag{1}$$

$$\tan R_1 = \sqrt{\left\{\frac{\cos(S-A)}{-\cos S\cos(S-B)\cos(S-C)}\right\}} = \frac{\cos(S-A)}{N}, \tag{2}$$

$$\tan R_1 = \frac{\sin\frac{1}{2}a}{\sin A\sin\frac{1}{2}b\sin\frac{1}{2}c}, \tag{3}$$

$$\tan R_1 = \frac{2\sin\frac{1}{2}a\cos\frac{1}{2}b\cos\frac{1}{2}c}{\sqrt{\{\sin s\sin(s-a)\sin(s-b)\sin(s-c)\}}}, \tag{4}$$

$$\tan R_1 = \frac{1}{2n}\{\sin s - \sin(s-a) + \sin(s-b) + \sin(s-c)\}. \tag{5}$$

Similarly we may find expressions for $\tan R_2$ and $\tan R_3$.

94. Many examples may be proposed involving properties of the circles inscribed in and described about the associated triangles. We will give one that will be of use hereafter.

To prove that

$$(\cot r + \tan R)^2 = \frac{1}{4n^2}(\sin a + \sin b + \sin c)^2 - 1.$$

We have

$$4n^2 = 1 - \cos^2 a - \cos^2 b - \cos^2 c + 2\cos a \cos b \cos c;$$

therefore

$$(\sin a + \sin b + \sin c)^2 - 4n^2$$

$= 2\ (1 + \sin a \sin b + \sin b \sin c + \sin c \sin a - \cos a \cos b \cos c\ ).$

Also $\cot r + \tan R = \dfrac{1}{2n}\left\{\sin s + \sin(s-a) + \sin(s-b) + \sin(s-c)\right\}$; and by squaring both members of this equation the required result will be obtained. For it may be shewn by reduction that

$$\sin^2 s + \sin^2(s-a) + \sin^2(s-b) + \sin^2(s-c) = 2 - 2\cos a \cos b \cos c,$$

and

$$\begin{gathered}\sin s \sin(s-a) + \sin s \sin(s-b) + \sin s \sin(s-c) \\ + \sin(s-a)\sin(s-b) + \sin(s-b)\sin(s-c) + \sin(s-c)\sin(s-a) \\ = \sin a \sin b + \sin b \sin c + \sin c \sin a.\end{gathered}$$

Similarly we may prove that

$$(\cot r_1 - \tan R)^2 = \frac{1}{4n^2}(\sin b + \sin c - \sin a)^2 - 1.$$

95. In the figure to Art. 89, suppose DP produced through P to a point A' such that DA' is a quadrant, then A' is a pole of BC, and $PA' = \dfrac{\pi}{2} - r$; similarly, suppose EP produced through P to a point B' such that EB' is a quadrant, and FP produced through P to a point C' such that FC' is a quadrant. Then $A'B'C'$ is the polar triangle of ABC, and $PA' = PB' = PC' = \dfrac{\pi}{2} - r$. Thus P is the pole of the small circle *described round* the polar triangle, and the angular radius of the small circle described round the polar triangle is the complement of the angular radius of the small circle inscribed in the primitive triangle. And in like manner the point which is the pole of the small circle inscribed in the polar triangle is also the pole of the small circle described round the primitive triangle, and the angular radii of the two circles are complementary.

EXAMPLES.

In the following examples the notation of the Chapter is retained. Shew that in any triangle the following relations hold contained in Examples 1 to 7:

1. $\text{Tan}\, r_1 \tan r_2 \tan r_3 = \tan r \sin^2 s$.

2. $$\text{Tan}\, R + \cot r = \tan R_1 + \cot r_1 = \tan R_2 + \cot r_2$$
$$= \tan R_3 + \cot r_3 = \tfrac{1}{2}(\cot r + \cot r_1 + \cot r_2 + \cot r_3).$$

3. $$\text{Tan}^2 R + \tan^2 R_1 + \tan^2 R_2 + \tan^2 R_3$$
$$= \cot^2 r + \cot^2 r_1 + \cot^2 r_2 + \cot^2 r_3.$$

4. $$\frac{\text{Tan}\, r_1 + \tan r_2 + \tan r_3 - \tan r}{\cot r_1 + \cot r_2 + \cot r_3 - \cot r} = \tfrac{1}{2}(1 + \cos a + \cos b + \cos c).$$

5. $\text{Cosec}^2 r = \cot(s-a)\cot(s-b) + \cot(s-b)\cot(s-c) + \cot(s-c)(s-a)$.

6. $\text{Cosec}^2 r_1 = \cot(s-b)\cot(s-c) - \cot s \cot(s-b) - \cot s \cot(s-c)$.

7. $\text{Tan}\, R_1 \tan R_2 \tan R_3 = \tan R \sec^2 S$.

8. Shew that in an equilateral triangle $\tan R = 2\tan r$.

9. If ABC be an equilateral spherical triangle, P the pole of the circle circumscribing it, Q any point on the sphere, shew that

$$\cos QA + \cos QB + \cos QC = 3\cos PA \cos PQ.$$

10. If three small circles be inscribed in a spherical triangle having each of its angles 120°, so that each touches the other two as well as two sides of the triangle, shew that the radius of each of the small circles = 30°, and that the centres of the three small circles coincide with the angular points of the polar triangle.

VIII

AREA OF A SPHERICAL TRIANGLE. SPHERICAL EXCESS.

96. *To find the area of a Lune.*

A *Lune* is that portion of the surface of a sphere which is comprised between two great semicircles.

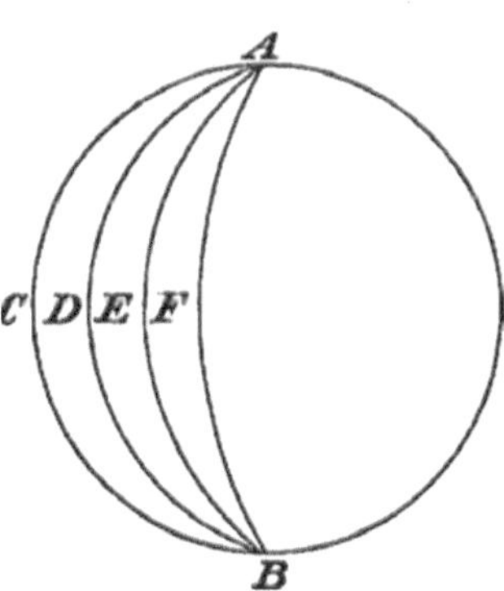

Let $ACBDA$, $ADBEA$ be two lunes having equal angles at A; then one of these lunes may be supposed placed on the other so as to coincide exactly with it; thus *lunes having equal angles are equal.* Then by a process similar to that used in the first proposition of the Sixth Book of Euclid it may be shewn that *lunes are proportional to their angles.* Hence since the whole surface of a sphere may be considered as a lune with an angle equal to four right angles, we have for a lune with an angle of which the circular measure is A,

$$\frac{\text{area of lune}}{\text{surface of sphere}} = \frac{A}{2\pi}.$$

Suppose r the radius of the sphere, then the surface is $4\pi r^2$ (*Integral Calculus*, Chap. VII.); thus

$$\text{area of lune} = \frac{A}{2\pi} 4\pi r^2 = 2Ar^2.$$

97. *To find the area of a Spherical Triangle.*

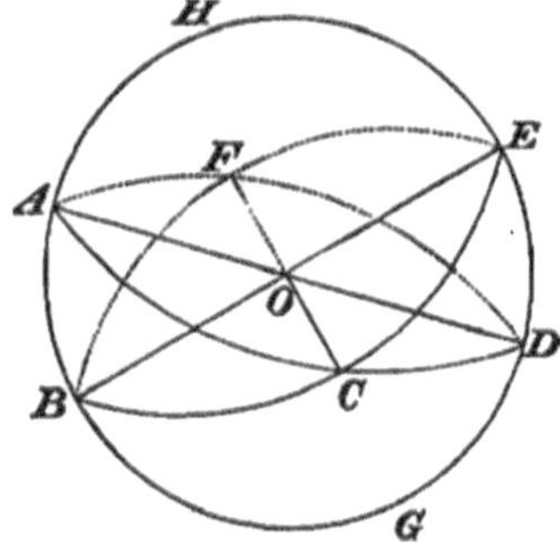

Let ABC be a spherical triangle; produce the arcs which form its sides until they meet again two and two, which will happen when each has become equal to the semicircumference. The triangle ABC now forms a part of three lunes, namely, $ABDCA$, $BCEAB$, and $CAFBC$. Now the triangles CDE and AFB are subtended by vertically opposite solid angles at O, and *we will assume* that their areas are equal; therefore the lune $CAFBC$ is equal to the sum of the two triangles ABC and CDE. Hence if A, B, C denote the circular measures of the angles of the triangle, we have

$$\text{triangle } ABC + BGDC = \text{lune } ABDCA = 2Ar^2,$$
$$\text{triangle } ABC + AHEC = \text{lune } BCEAB = 2Br^2,$$
$$\text{triangle } ABC + \text{triangle } CDE = \text{lune } CAFBC = 2Cr^2;$$

hence, by addition,

$$\text{twice triangle } ABC + \text{surface of hemisphere} = 2(A+B+C)r^2;$$

therefore $$\text{triangle } ABC = (A+B+C-\pi)r^2.$$

The expression $A+B+C-\pi$ is called the *spherical excess* of the triangle; and since

$$(A+B+C-\pi)r^2 = \frac{A+B+C-\pi}{2\pi}\,2\pi r^2,$$

the result obtained may be thus enunciated: *the area of a spherical triangle is the same fraction of half the surface of the sphere as the spherical excess is of four right angles.*

98. We have assumed, as is usually done, that the areas of the triangles CDE and AFB in the preceding Article are equal. The triangles are, however, not absolutely equal, but *symmetrically* equal (Art. 57), so that one cannot be made to coincide with the other by superposition. It is, however, easy to decompose two such triangles into pieces which admit of superposition, and thus to prove that their areas are equal. For describe a small circle round each, then the angular radii of these circles will be equal by Art. 92. If the pole of the circumscribing circle falls inside each triangle, then each triangle is the sum of three isosceles triangles, and if the pole falls outside each triangle, then each triangle is the excess of two isosceles triangles over a third; and in each case the isosceles triangles of one set are respectively *absolutely equal* to the corresponding isosceles triangles of the other set.

99. *To find the area of a spherical polygon.*

Let n be the number of sides of the polygon, Σ the sum of all its angles. Take any point within the polygon and join it with all the angular points; thus the figure is divided into n triangles. Hence, by Art. 97,

$$\text{area of polygon} = (\text{sum of the angles of the triangles} - n\pi)r^2,$$

and the sum of the angles of the triangles is equal to Σ together with the four right angles which are formed round the common vertex; therefore

$$\text{area of polygon} = \left\{\Sigma - (n-2)\pi\right\}r^2.$$

This expression is true even when the polygon has some of its angles greater than two right angles, provided it can be decomposed into triangles, of which each of the angles is less than two right angles.

100. We shall now give some expressions for certain trigonometrical functions of the *spherical excess* of a triangle. We denote the spherical excess by E, so that $E = A + B + C - \pi$.

101. *Cagnoli's Theorem.* To shew that

$$\sin \tfrac{1}{2}E = \frac{\sqrt{\{\sin s \sin(s-a)\sin(s-b)\sin(s-c)\}x}}{2\cos \tfrac{1}{2}a \cos \tfrac{1}{2}b \cos \tfrac{1}{2}c}.$$

$$\begin{aligned}
\text{Sin}\, \tfrac{1}{2}E &= \sin \tfrac{1}{2}(A+B+C-\pi) = \sin\{\tfrac{1}{2}(A+B) - \tfrac{1}{2}(\pi - C)\} \\
&= \sin \tfrac{1}{2}(A+B)\sin \tfrac{1}{2}C - \cos \tfrac{1}{2}(A+B)\cos \tfrac{1}{2}C \\
&= \frac{\sin \tfrac{1}{2}C \cos \tfrac{1}{2}C}{\cos \tfrac{1}{2}c}\{\cos \tfrac{1}{2}(a-b) - \cos \tfrac{1}{2}(a-b)\}, \quad \text{(Art. 54)}, \\
&= \frac{\sin C \sin \tfrac{1}{2}a \sin \tfrac{1}{2}b}{\cos \tfrac{1}{2}c} \\
&= \frac{\sin \tfrac{1}{2}a \sin \tfrac{1}{2}b}{\cos \tfrac{1}{2}c} \cdot \frac{2}{\sin a \sin b} \cdot \sqrt{\{\sin s \sin(s-a)\sin(s-b)\sin(s-c)\}} \\
&= \frac{\sqrt{\{\sin s \sin(s-a)\sin(s-b)\sin(s-c)\}}}{2\cos \tfrac{1}{2}a \cos \tfrac{1}{2}b \cos \tfrac{1}{2}c}.
\end{aligned}$$

102. *Lhuilier's Theorem.* To shew that

$$\tan \tfrac{1}{4}E = \sqrt{\{\tan \tfrac{1}{2}s \tan \tfrac{1}{2}(s-a)\tan \tfrac{1}{2}(s-b)\tan \tfrac{1}{2}(s-c)\}}.$$

$$\begin{aligned}
\text{Tan}\, \tfrac{1}{4}E &= \frac{\sin \tfrac{1}{2}(A+B+C-\pi)}{\cos \tfrac{1}{4}(A+B+C-\pi)} \\
&= \frac{\sin \tfrac{1}{2}(A+B) - \sin \tfrac{1}{2}(\pi - C)}{\cos \tfrac{1}{2}(A+B) + \cos \tfrac{1}{2}(\pi - C)}, \qquad (\textit{Plane Trig.}\ \text{Art. } 84), \\
&= \frac{\sin \tfrac{1}{2}(A+B) - \cos \tfrac{1}{2}C}{\cos \tfrac{1}{2}(A+B) + \sin \tfrac{1}{2}C}
\end{aligned}$$

$$= \frac{\cos\frac{1}{2}(a-b) - \cos\frac{1}{2}c}{\cos\frac{1}{2}(a+b) + \cos\frac{1}{2}c} \cdot \frac{\cos\frac{1}{2}C}{\sin\frac{1}{2}C}, \qquad \text{(Art. 54).}$$

Hence, by Art. 45, we obtain

$$\tan\tfrac{1}{4}E = \frac{\sin\frac{1}{4}(c+a-b)\sin\frac{1}{4}(c+b-a)}{\cos\frac{1}{4}(a+b+c)\cos\frac{1}{4}(a+b-c)}\sqrt{\left\{\frac{\sin s\sin(s-c)}{\sin(s-a)\sin(s-b)}\right\}}$$

$$= \sqrt{\{\tan\tfrac{1}{2}s\tan\tfrac{1}{2}(s-a)\tan\tfrac{1}{2}(s-b)\tan\tfrac{1}{2}(s-c)\}}.$$

103. We may obtain many other formulæ involving trigonometrical functions of the spherical excess. Thus, for example,

$$\cos\tfrac{1}{2}E = \cos\left\{\tfrac{1}{2}(A+B) - \tfrac{1}{2}(\pi - C)\right\}$$

$$= \cos\tfrac{1}{2}(A+B)\sin\tfrac{1}{2}C + \sin\tfrac{1}{2}(A+B)\cos\tfrac{1}{2}C$$

$$= \left\{\cos\tfrac{1}{2}(a+b)\sin^2\tfrac{1}{2}C + \cos\tfrac{1}{2}(a-b)\cos^2\tfrac{1}{2}C\right\}\sec\tfrac{1}{2}c, \text{ (Art. 54)},$$

$$= \Big\{\cos\tfrac{1}{2}a\cos\tfrac{1}{2}b(\cos^2 C + \sin^2\tfrac{1}{2}C) + \sin\tfrac{1}{2}a\sin\tfrac{1}{2}b(\cos^2\tfrac{1}{2}C - \sin^2\tfrac{1}{2}C)\Big\}\sec\tfrac{1}{2}c$$

$$= \left\{\cos\tfrac{1}{2}a\cos\tfrac{1}{2}b + \sin\tfrac{1}{2}a\sin\tfrac{1}{2}b\cos C\right\}\sec\tfrac{1}{2}c. \qquad (1)$$

Again, it was shewn in Art. 101, that

$$\sin\tfrac{1}{2}E = \sin C\sin\tfrac{1}{2}a\sin\tfrac{1}{2}b\sec\tfrac{1}{2}c;$$

therefore

$$\tan\tfrac{1}{2}E = \frac{\sin\frac{1}{2}a\sin\frac{1}{2}b\sin C}{\cos\frac{1}{2}a\cos\frac{1}{2}b + \sin\frac{1}{2}a\sin\frac{1}{2}b\cos C}. \qquad (2)$$

Again, we have from above

$$\cos\tfrac{1}{2}E = \left\{\cos\tfrac{1}{2}a\cos\tfrac{1}{2}b + \sin\tfrac{1}{2}a\sin\tfrac{1}{2}b\cos C\right\}\sec\tfrac{1}{2}c$$

$$= \frac{(1+\cos a)(1+\cos b) + \sin a\sin b\cos C}{4\cos\frac{1}{2}a\cos\frac{1}{2}b\cos\frac{1}{2}c}$$

$$= \frac{1 + \cos a + \cos b \cos c}{4 \cos \frac{1}{2}a \cos \frac{1}{2}b \cos \frac{1}{2}c} = \frac{\cos^2 \frac{1}{2}a + \cos^2 \frac{1}{2}b + \cos^2 \frac{1}{2}c - 1}{2 \cos \frac{1}{2}a \cos \frac{1}{2}b \cos \frac{1}{2}c}. \tag{3}$$

In (3) put $1 - 2\sin^2 \frac{1}{4}E$ for $\cos \frac{1}{2}E$; thus

$$\sin^2 \tfrac{1}{4}E = \frac{1 + 2\cos \frac{1}{2}a \cos \frac{1}{2}b \cos \frac{1}{2}c - \cos^2 \frac{1}{2}a - \cos^2 \frac{1}{2}b - \cos^2 \frac{1}{2}c}{4 \cos \frac{1}{2}a \cos \frac{1}{2}b \cos \frac{1}{2}c}.$$

By ordinary development we can shew that the numerator of the above fraction is equal to

$$4 \sin \tfrac{1}{2}s \sin \tfrac{1}{2}(s-a) \sin \tfrac{1}{2}(s-b) \sin \tfrac{1}{2}(s-c);$$

therefore

$$\sin^2 \tfrac{1}{4}E = \frac{\sin \frac{1}{2}s \sin \frac{1}{2}(s-a) \sin \frac{1}{2}(s-b) \sin \frac{1}{2}(s-c)}{\cos \frac{1}{2}a \cos \frac{1}{2}b \cos \frac{1}{2}c}. \tag{4}$$

Similarly

$$\cos^2 \tfrac{1}{4}E = \frac{\cos \frac{1}{2}s \cos \frac{1}{2}(s-a) \cos \frac{1}{2}(s-b) \cos \frac{1}{2}(s-c)}{\cos \frac{1}{2}a \cos \frac{1}{2}b \cos \frac{1}{2}c}. \tag{5}$$

Hence by division we obtain Lhuilier's Theorem.

Again,

$$\begin{aligned}\frac{\sin(C - \frac{1}{2}E)}{\sin \frac{1}{2}E} &= \sin C \cot \tfrac{1}{2}E - \cos C \\ &= \sin C \frac{\cos \frac{1}{2}a \cos \frac{1}{2}b + \sin \frac{1}{2}a \sin \frac{1}{2}b \cos C}{\sin \frac{1}{2}a \sin \frac{1}{2}b \sin C} - \cos C, \text{ by (2)}, \\ &= \cos \tfrac{1}{2}a \cot \tfrac{1}{2}b;\end{aligned}$$

therefore, by Art. 101,

$$\sin(C - \tfrac{1}{2}E) = \frac{\sqrt{\{\sin s \sin(s-a) \sin(s-b) \sin(s-c)\}}}{2 \sin \frac{1}{2}a \sin \frac{1}{2}b \cos \frac{1}{2}c}.$$

Again, $\cos(C - \frac{1}{2}E) = \cos C \cos \frac{1}{2}E + \sin C \sin \frac{1}{2}E$

$$= \frac{(1 + \cos a)(1 + \cos b)\cos C + \sin a \sin b \cos^2 C}{4 \cos \frac{1}{2}a \cos \frac{1}{2}b \cos \frac{1}{2}c} + \sin^2 C \sin \frac{1}{2}a \sin \frac{1}{2}b \sec \frac{1}{2}c$$

$$= \frac{(1 + \cos a)(1 + \cos b)\cos C + \sin a \sin b}{4 \cos \frac{1}{2}a \cos \frac{1}{2}b \cos \frac{1}{2}c}$$

$$= \left\{\cos \tfrac{1}{2}a \cos \tfrac{1}{2}b \cos C + \sin \tfrac{1}{2}a \sin \tfrac{1}{2}b\right\} \sec \tfrac{1}{2}c$$

$$= \frac{\sin a \sin b \cos C + 4 \sin^2 \frac{1}{2}a \sin^2 \frac{1}{2}b}{4 \sin \frac{1}{2}a \sin \frac{1}{2}b \cos \frac{1}{2}c}$$

$$= \frac{\cos c - \cos a \cos b + (1 - \cos a)(1 - \cos b)}{4 \sin \frac{1}{2}a \sin \frac{1}{2}b \cos \frac{1}{2}c}$$

$$= \frac{1 + \cos c - \cos a - \cos b}{4 \sin \frac{1}{2}a \sin \frac{1}{2}b \cos \frac{1}{2}c} = \frac{\cos^2 \frac{1}{2}c - \cos^2 \frac{1}{2}a - \cos^2 \frac{1}{2}b + 1}{2 \sin \frac{1}{2}a \sin \frac{1}{2}b \cos \frac{1}{2}c}. \quad (6)$$

From this result we can deduce two other results, in the same manner as (4) and (5) were deduced from (3); or we may observe that the right-hand member of (6) can be obtained from the right-hand member of (3) by writing $\pi - a$ and $\pi - b$ for a and b respectively, and thus we may deduce the results more easily. We shall have then

$$\sin^2(\tfrac{1}{2}C - \tfrac{1}{4}E) = \frac{\cos \frac{1}{2}s \sin \frac{1}{2}(s - a) \sin \frac{1}{2}(s - b) \cos \frac{1}{2}(s - c)}{\sin \frac{1}{2}a \sin \frac{1}{2}b \cos \frac{1}{2}c},$$

$$\cos^2(\tfrac{1}{2}C - \tfrac{1}{4}E) = \frac{\sin \frac{1}{2}s \cos \frac{1}{2}(s - a) \cos \frac{1}{2}(s - b) \sin \frac{1}{2}(s - c)}{\sin \frac{1}{2}a \sin \frac{1}{2}b \cos \frac{1}{2}c}.$$

EXAMPLES.

1. Find the angles and sides of an equilateral triangle whose area is one-fourth of that of the sphere on which it is described.

2. Find the surface of an equilateral and equiangular spherical polygon of n sides, and determine the value of each of the angles when the surface equals half the surface of the sphere.

3. If $a = b = \dfrac{\pi}{3}$, and $c = \dfrac{\pi}{2}$, shew that $E = \cos^{-1} \dfrac{7}{9}$.

4. If the angle C of a spherical triangle be a right angle, shew that

$$\sin\tfrac{1}{2}E = \sin\tfrac{1}{2}a\sin\tfrac{1}{2}b\sec\tfrac{1}{2}c, \quad \cos\tfrac{1}{2}E = \cos\tfrac{1}{2}a\cos\tfrac{1}{2}b\sec\tfrac{1}{2}c.$$

5. If the angle C be a right angle, shew that

$$\frac{\sin^2 c}{\cos c}\cos E = \frac{\sin^2 a}{\cos a} + \frac{\sin^2 b}{\cos b}.$$

6. If $a = b$ and $C = \dfrac{\pi}{2}$, shew that $\tan E = \dfrac{\sin^2 a}{2\cos a}$.

7. The sum of the angles in a right-angled triangle is less than four right angles.

8. Draw through a given point in the side of a spherical triangle an arc of a great circle cutting off a given part of the triangle.

9. In a spherical triangle if $\cos C = -\tan\dfrac{a}{2}\tan\dfrac{b}{2}$, then $C = A + B$.

10. If the angles of a spherical triangle be together equal to four right angles

$$\cos^2\tfrac{1}{2}a + \cos^2\tfrac{1}{2}b + \cos^2\tfrac{1}{2}c = 1.$$

11. If r_1, r_2, r_3 be the radii of three small circles of a sphere of radius r which touch one another at P, Q, R, and A, B, C be the angles of the spherical triangle formed by joining their centres,

$$\text{area } PQR = (A\cos r_1 + B\cos r_2 + C\cos r_3 - \pi)r^2.$$

12. Shew that

$$\sin s = \frac{\left\{\sin\frac{1}{2}E\sin(A-\frac{1}{2}E)\sin(B-\frac{1}{2}E)\sin(C-\frac{1}{2}E)\right\}^{\frac{1}{2}}}{2\sin\frac{1}{2}A\sin\frac{1}{2}B\sin\frac{1}{2}C}.$$

13. Given two sides of a spherical triangle, determine when the area is a maximum.

14. Find the area of a regular polygon of a given number of sides formed by arcs of great circles on the surface of a sphere; and hence deduce that, if α be the angular radius of a small circle, its area is to that of the whole surface of the sphere as $\operatorname{versin}\alpha$ is to 2.

15. A, B, C are the angular points of a spherical triangle; A', B', C' are the middle points of the respectively opposite sides. If E be the spherical excess of the triangle, shew that

$$\cos \tfrac{1}{2}E = \frac{\cos A'B'}{\cos \frac{1}{2}c} = \frac{\cos B'C'}{\cos \frac{1}{2}a} = \frac{\cos C'A'}{\cos \frac{1}{2}b}.$$

16. If one of the arcs of great circles which join the middle points of the sides of a spherical triangle be a quadrant, shew that the other two are also quadrants.

IX

ON CERTAIN APPROXIMATE FORMULÆ.

104. We shall now investigate certain approximate formulæ which are often useful in calculating spherical triangles when the radius of the sphere is large compared with the lengths of the sides of the triangles.

105. *Given two sides and the included angle of a spherical triangle, to find the angle between the chords of these sides.*

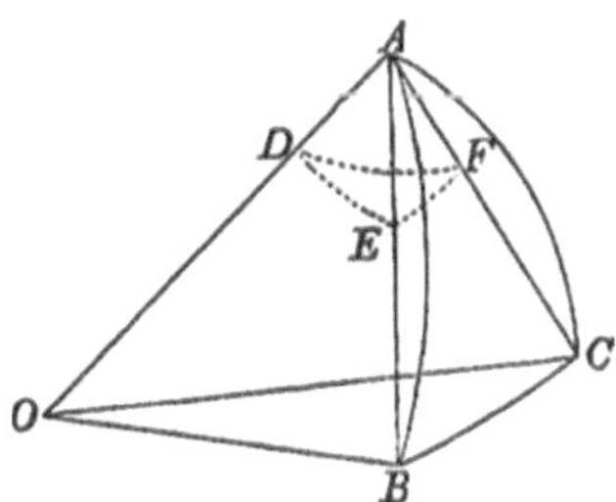

Let AB, AC be the two sides of the triangle ABC; let O be the centre of the sphere. Describe a sphere round A as a centre, and suppose it to meet AO, AB, AC at D, E, F respectively. Then the angle EDF is the inclination of the planes OAB, OAC, and is therefore equal to A. From the spherical triangle DEF

$$\cos EF = \cos DE \cos DF + \sin DE \sin DF \cos A;$$

and $$DE = \tfrac{1}{2}(\pi - c), \quad DF = \tfrac{1}{2}(\pi - b);$$

therefore $$\cos EF = \sin \tfrac{1}{2}b \sin \tfrac{1}{2}c + \cos \tfrac{1}{2}b \cos \tfrac{1}{2}c \cos A.$$

If the sides of the triangle are small compared with the radius of the sphere, EF will not differ much from A; suppose $EF = A - \theta$, then approximately

$$\cos EF = \cos A + \theta \sin A;$$

and $$\sin \tfrac{1}{2}b \sin \tfrac{1}{2}c = \sin^2 \tfrac{1}{4}(b + c) - \sin^2 \tfrac{1}{4}(b - c),$$

$$\cos \tfrac{1}{2}b \cos \tfrac{1}{2}c = \cos^2 \tfrac{1}{4}(b + c) - \sin^2 \tfrac{1}{4}(b - c);$$

therefore

$$\cos A + \theta \sin A = \sin^2 \tfrac{1}{4}(b + c) - \sin^2 \tfrac{1}{4}(b - c) + \left\{1 - \sin^2 \tfrac{1}{4}(b + c) - \sin^2 \tfrac{1}{4}(b - c)\right\} \cos A;$$

therefore

$$\theta \sin A = (1 - \cos A) \sin^2 \tfrac{1}{4}(b + c) - (1 + \cos A) \sin^2 \tfrac{1}{4}(b - c),$$

therefore $$\theta = \tan \tfrac{1}{2}A \sin^2 \tfrac{1}{4}(b + c) - \cot \tfrac{1}{2}A \sin^2 \tfrac{1}{4}(b - c).$$

This gives the *circular measure* of θ; the number of seconds in the angle is found by dividing the circular measure by the circular measure of one second, or approximately by the sine of one second (*Plane Trigonometry*, Art. 123). If the lengths of the arcs corresponding to a and b respectively be α and β, and r the radius of the sphere, we have $\frac{\alpha}{r}$ and $\frac{\beta}{r}$ as the circular measures of a and b respectively; and the lengths of the sides of the chordal triangle are $2r \sin \frac{\alpha}{2r}$ and $2r \sin \frac{\beta}{2r}$ respectively. Thus when the sides of the spherical triangle and the radius of the sphere are known, we can calculate the angles and sides of the chordal triangle.

106. Legendre's Theorem. *If the sides of a spherical triangle be small compared with the radius of the sphere, then each angle of the spherical triangle exceeds by one third of the spherical excess the corresponding angle of the plane triangle, the sides of which are of the same length as the arcs of the spherical triangle.*

Let A, B, C be the angles of the spherical triangle; a, b, c the sides; r the radius of the sphere; α, β, γ the lengths of the arcs which form the sides, so that $\frac{\alpha}{r}, \frac{\beta}{r}, \frac{\gamma}{r}$ are the circular measures of a, b, c respectively. Then

$$\cos A = \frac{\cos a - \cos b \cos c}{\sin b \sin c};$$

now

$$\cos a = 1 - \frac{\alpha^2}{2r^2} + \frac{\alpha^4}{24r^4} - \dots,$$

$$\sin a = \frac{\alpha}{r} - \frac{\alpha^3}{6r^3} + \dots.$$

Similar expressions hold for $\cos b$ and $\sin b$, and for $\cos c$ and $\sin c$ respectively. Hence, if we neglect powers of the circular measure above the *fourth*, we have

$$\cos A = \frac{1 - \frac{\alpha^2}{2r^2} + \frac{\alpha^4}{24r^4} - \left(1 - \frac{\beta^2}{2r^2} + \frac{\beta^4}{24r^4}\right)\left(1 - \frac{\gamma^2}{2r^2} + \frac{\gamma^4}{24r^4}\right)}{\frac{\beta\gamma}{r^2}\left(1 - \frac{\beta^2}{6r^2}\right)\left(1 - \frac{\gamma^2}{6r^2}\right)}$$

$$= \frac{\frac{1}{2r^2}(\beta^2 + \gamma^2 - \alpha^2) + \frac{1}{24r^4}(\alpha^4 - \beta^4 - \gamma^4 - 6\beta^2\gamma^2)}{\frac{\beta\gamma}{r^2}\left(1 - \frac{\beta^2 + \gamma^2}{6r^2}\right)}$$

$$= \frac{1}{2\beta\gamma}\left\{\beta^2 + \gamma^2 - \alpha^2 + \frac{1}{12r^2}(\alpha^2 - \beta^2 - \gamma^2 - 6\beta^2\gamma^2)\right\}\left\{1 + \frac{\beta^2 + \gamma^2}{6r^2}\right\}$$

$$= \frac{\beta^2 + \gamma^2 - \alpha^2}{2\beta\gamma} + \frac{\alpha^4 + \beta^4 + \gamma^4 - 2\alpha^2\beta^2 - 2\beta^2\gamma^2 - 2\gamma^2\alpha^2}{24\beta\gamma r^2}.$$

Now let A', B', C' be the angles of the plane triangle whose sides are α, β, γ respectively; then

$$\cos A' = \frac{\beta^2 + \gamma^2 - \alpha^2}{2\beta\gamma},$$

thus

$$\cos A = \cos A' - \frac{\beta\gamma \sin^2 A'}{6r^2}.$$

Suppose $A = A' + \theta$; then

$$\cos A = \cos A' - \theta \sin A' \text{ approximately};$$

therefore
$$\theta = \frac{\beta\gamma \sin A'}{6r^2} = \frac{S}{3r^2},$$
where S denotes the area of the plane triangle whose sides are α, β, γ. Similarly

$$B = B' + \frac{S}{3r^2} \text{ and } C = C' + \frac{S}{3r^2};$$

hence approximately

$$A + B + C = A' + B' + C' + \frac{S}{r^2} = \pi + \frac{S}{r^2};$$

therefore $\frac{S}{r^2}$ is approximately equal to the spherical excess of the spherical triangle, and thus the theorem is established.

It will be seen that in the above approximation the area of the spherical triangle is considered equal to the area of the plane triangle which can be formed with sides of the same length.

107. Legendre's Theorem may be used for the approximate solution of spherical triangles in the following manner.

(1) Suppose the three sides of a spherical triangle known; then the values of α, β, γ are known, and by the formulæ of Plane Trigonometry we can calculate S and A', B', C'; then A, B, C are known from the formulæ.

$$A = A' + \frac{S}{3r^2}, \quad B = B' + \frac{S}{3r^2}, \quad C = C' + \frac{S}{3r^2}.$$

(2) Suppose two sides and the included angle of a spherical triangle known, for example A, b, c. Then

$$S = \tfrac{1}{2}\beta\gamma \sin A' = \tfrac{1}{2}\beta\gamma \sin A \text{ approximately}.$$

Then A' is known from the formula $A' = A - \frac{S}{3r^2}$. Thus in the plane triangle two sides and the included angle are known; therefore its remaining parts can be calculated, and then those of the spherical triangle become known.

(3) Suppose two sides and the angle opposite to one of them in a spherical

triangle known, for example A, a, b. Then

$$\sin B' = \frac{\beta}{\alpha} \sin A' = \frac{\beta}{\alpha} \sin A \text{ approximately;}$$

and $C' = \pi - A' - B' = \pi - A - B'$ approximately; then $S = \frac{1}{2}\alpha\beta \sin C'$. Hence A' is known and the plane triangle can be solved, since two sides and the angle opposite to one of them are known.

(4) Suppose two angles and the included side of a spherical triangle known, for example A, B, c.

$$\text{Then } S = \frac{\gamma^2 \sin A' \sin B'}{2\sin(A' + B')} = \frac{\gamma^2 \sin A \sin B}{2\sin(A + B)} \text{ nearly.}$$

Hence in the plane triangle two angles and the included side are known.

(5) Suppose two angles and the side opposite to one of them in a spherical triangle known, for example A, B, a. Then

$$C' = \pi - A' - B' = \pi - A - B, \text{ approximately, and}$$

$$S = \frac{\alpha^2 \sin B' \sin C'}{2\sin(B' + C')},$$

which can be calculated, since B' and C' are approximately known.

108. The importance of Legendre's Theorem in the application of Spherical Trigonometry to the measurement of the Earth's surface has given rise to various developments of it which enable us to test the degree of exactness of the approximation. We shall finish the present Chapter with some of these developments, which will serve as exercises for the student. We have seen that approximately the spherical excess is equal to $\frac{S}{r^2}$, and we shall begin with investigating a closer approximate formula for the spherical excess.

109. *To find an approximate value of the spherical excess.*

Let E denote the spherical excess; then

$$\sin \frac{1}{2} E = \frac{\sin \frac{1}{2}a \sin \frac{1}{2}b \sin C}{\cos \frac{1}{2}c};$$

therefore approximately

$$\sin \tfrac{1}{2}E = \sin C \frac{\alpha\beta}{4r^2}\left(1 - \frac{\alpha^2}{24r^2}\right)\left(1 - \frac{\beta^2}{24r^2}\right)\left(1 - \frac{\gamma^2}{8r^2}\right)^{-1}$$

$$= \sin C \frac{\alpha\beta}{4r^2}\left(1 + \frac{3\gamma^2 - \alpha^2 - \beta^2}{24r^2}\right);$$

therefore
$$E = \sin C \frac{\alpha\beta}{2r^2}\left(1 + \frac{3\gamma^2 - \alpha^2 - \beta^2}{24r^2}\right), \qquad (1)$$

and
$$\sin C = \sin\left(C' + \tfrac{1}{3}E\right) = \sin C' + \tfrac{1}{3}E\cos C'$$

$$= \sin C' + \frac{\sin C' \cos C'}{3}\frac{\alpha\beta}{2r^2} = \sin C'\left(1 + \frac{\alpha^2 + \beta^2 - \gamma^2}{12r^2}\right). \qquad (2)$$

From (1) and (2)

$$E = \sin C' \frac{\alpha\beta}{2r^2}\left(1 + \frac{\alpha^2 + \beta^2 + \gamma^2}{24r^2}\right).$$

Hence to this order of approximation the area of the spherical triangle exceeds that of the plane triangle by the fraction $\dfrac{\alpha^2 + \beta^2 + \gamma^2}{24r^2}$ of the latter.

110. *To find an approximate value of* $\dfrac{\sin A}{\sin B}$.

$$\frac{\operatorname{Sin} A}{\operatorname{Sin} B} = \frac{\sin a}{\sin b};$$

hence approximately
$$\frac{\sin A}{\sin B} = \frac{\alpha\left(1 - \dfrac{\alpha^2}{6r^2} + \dfrac{\alpha^4}{120r^4}\right)}{\beta\left(1 - \dfrac{\beta^2}{6r^2} + \dfrac{\beta^4}{120r^4}\right)}$$

$$= \frac{\alpha}{\beta}\left(1 - \frac{\alpha^2}{6r^2} + \frac{\alpha^4}{120r^4} + \frac{\beta^2}{6r^2} - \frac{\alpha^2\beta^2}{36r^4} - \frac{\beta^4}{120r^4} + \frac{\beta^4}{36r^4}\right)$$

$$= \frac{\alpha}{\beta}\left\{1 + \frac{\beta^2 - \alpha^2}{6r^2} + \frac{\alpha^4 - \beta^4}{120r^4} + \frac{\beta^2(\beta^2 - \alpha^2)}{36r^4}\right\}$$

$$= \frac{\alpha}{\beta}\left\{1 + \frac{\beta^2 - \alpha^2}{6r^2}\left(1 + \frac{\beta^2}{6r^2} - \frac{\alpha^2 + \beta^2}{20r^2}\right)\right\}$$

$$= \frac{\alpha}{\beta}\left\{1 + \frac{\beta^2 - \alpha^2}{6r^2}\left(1 + \frac{7\beta^2 - 3\alpha^2}{60r^2}\right)\right\}.$$

111. *To express* $\cot B - \cot A$ *approximately.*

$$\text{Cot}\, B - \cot A = \frac{1}{\sin B}(\cos B - \frac{\sin B}{\sin A}\cos A);$$

hence, approximately, by Art. 110,

$$\cot B - \cot A = \frac{1}{\sin B}(\cos B - \frac{\beta}{\alpha}\cos A - \frac{\beta}{\alpha}\frac{\alpha^2 - \beta^2}{6r^2}\cos A).$$

Now we have shewn in Art. 106, that approximately

$$\cos A = \frac{\beta^2 + \gamma^2 - \alpha^2}{2\beta\gamma} + \frac{\alpha^4 + \beta^4 + \gamma^4 - 2\alpha^2\beta^2 - 2\beta^2\gamma^2 - 2\gamma^2\alpha^2}{24\beta\gamma r^2},$$

therefore $\quad \cos B - \frac{\beta}{\alpha}\cos A = \frac{\alpha^2 - \beta^2}{\alpha\gamma}$ approximately,

and

$$\cot B - \cot A = \frac{\alpha^2 - \beta^2}{\alpha\gamma \sin B} - \frac{\alpha^2 - \beta^2}{\alpha\gamma\sin B}\frac{\beta^2 + \gamma^2 - \alpha^2}{12r^2}$$

$$= \frac{\alpha^2 - \beta^2}{\alpha\gamma\sin B}\left(1 - \frac{\beta^2 + \gamma^2 - \alpha^2}{12r^2}\right).$$

112. The approximations in Arts. 109 and 110 are true so far as terms involving r^4; that in Art. 111 is true so far as terms involving r^2, and it will be seen that we are thus able to carry the approximations in the following Article so far as terms involving r^4.

113. *To find an approximate value of the error in the length of a side of a spherical triangle when calculated by Legendre's Theorem.*

Suppose the side β known and the side α required; let 3μ denote the spherical excess which is adopted. Then the approximate value $\dfrac{\beta\sin(A - \mu)}{\sin(B - \mu)}$ is taken for the side of which α is the real value. Let $x = \alpha - \dfrac{\beta(A - \mu)}{\sin(B - \mu)}$; we have then to find x approximately. Now approximately

$$\frac{\sin(A-\mu)}{\sin(B-\mu)} = \frac{\sin A - \mu\cos A - \dfrac{\mu^2}{2}\sin A}{\sin B - \mu\cos B - \dfrac{\mu^2}{2}\sin B}$$

$$= \frac{\sin A}{\sin B}\left(1 - \mu\cot A - \frac{\mu^2}{2}\right)\left(1 - \mu\cot B - \frac{\mu^2}{2}\right)^{-1}$$
$$= \frac{\sin A}{\sin B}\left\{1 + \mu(\cot B - \cot A) + \mu^2\cot B(\cot B - \cot A)\right\}$$
$$= \frac{\sin A}{\sin B} + \frac{\mu\sin A}{\sin B}(\cot B - \cot A)(1 + \mu\cot B).$$

Also the following formulæ are true so far as terms involving r^2 :

$$\frac{\sin A}{\sin B} = \frac{\alpha}{\beta}\left(1 + \frac{\beta^2 - \alpha^2}{6r^2}\right),$$

$$\cot B - \cot A = \frac{\alpha^2 - \beta^2}{\alpha\gamma\sin B}\left(1 - \frac{\beta^2 + \gamma^2 - \alpha^2}{12r^2}\right),$$

$$1 + \mu\cot B = 1 + \frac{\alpha^2 + \gamma^2 - \beta^2}{12r^2}.$$

Hence, approximately,

$$\frac{\sin A}{\sin B}(\cot B - \cot A)(1 + \mu\cot B) = \frac{\alpha^2 - \beta^2}{\beta\gamma\sin B}.$$

Therefore $$x = \alpha - \frac{\beta\sin A}{\sin B} - \frac{\mu(\alpha^2 - \beta^2)}{\gamma\sin B}$$

$$= \frac{\alpha(\beta^2 - \alpha^2)}{6}\left\{\frac{6\mu}{\alpha\gamma\sin B} - \frac{1}{r^2} + \frac{3\alpha^2 - 7\beta^2}{60r^4}\right\},\text{ by Art. 110.}$$

If we calculate μ from the formula $\mu = \dfrac{\alpha\gamma\sin B}{6r^2}$ we obtain

$$x = \frac{\alpha(\beta^2 - \alpha^2)(3\alpha^2 - 7\beta^2)}{360r^4}.$$

If we calculate μ from an equation corresponding to (1) of Art. 109, we have

$$\mu = \frac{\alpha\gamma \sin B}{6r^2}\left(1 + \frac{3\beta^2 - \alpha^2 - \gamma^2}{24r^2}\right);$$

therefore
$$x = \frac{\alpha(\beta^2 - \alpha^2)(\alpha^2 + \beta^2 - 5\gamma^2)}{720r^4}.$$

MISCELLANEOUS EXAMPLES.

1. If the sides of a spherical triangle AB, AC be produced to B', C', so that BB', CC' are the semi-supplements of AB, AC respectively, shew that the arc $B'C'$ will subtend an angle at the centre of the sphere equal to the angle between the chords of AB and AC.

2. Deduce Legendre's Theorem from the formula

$$\tan^2 \frac{A}{2} = \frac{\sin \frac{1}{2}(a+b-c)\sin\frac{1}{2}(c+a-b)}{\sin\frac{1}{2}(b+c-a)\sin\frac{1}{2}(a+b+c)}.$$

3. Four points A, B, C, D on the surface of a sphere are joined by arcs of great circles, and E, F are the middle points of the arcs AC, BD: shew that

$$\cos AB + \cos BC + \cos CD + \cos DA = 4 \cos AE \cos BF \cos FE.$$

4. If a quadrilateral $ABCD$ be inscribed in a small circle on a sphere so that two opposite angles A and C may be at opposite extremities of a diameter, the sum of the cosines of the sides is constant.

5. In a spherical triangle if $A = B = 2C$, shew that

$$\cos a \cos \frac{a}{2} = \cos\left(c + \frac{a}{2}\right).$$

6. ABC is a spherical triangle each of whose sides is a quadrant; P is any point within the triangle: shew that

$$\cos PA \cos PB \cos PC + \cot BPC \cot CPA \cot APB = 0,$$

and
$$\tan ABP \tan BCP \tan CAP = 1.$$

7. If O be the middle point of an equilateral triangle ABC, and P any point

on the surface of the sphere, then

$$\tfrac{1}{4}(\tan PO \tan OA)^2 (\cos PA + \cos PB + \cos PC)^2 =$$
$$\cos^2 PA + \cos^2 PB + \cos^2 PC - \cos PA \cos PB - \cos PB \cos PC - \cos PC \cos PA.$$

8. If ABC be a triangle having each side a quadrant, O the pole of the inscribed circle, P any point on the sphere, then

$$(\cos PA + \cos PB + \cos PC)^2 = 3\cos^2 PO.$$

9. From each of three points on the surface of a sphere arcs are drawn on the surface to three other points situated on a great circle of the sphere, and their cosines are a, b, c; a', b', c'; a'', b'', c''. Shew that $ab''c' + a'bc'' + a''b'c = ab'c'' + a'b''c + a''bc'$.

10. From Arts. 110 and 111, shew that approximately

$$\log\beta = \log\alpha + \log\sin B - \log\sin A + \frac{S}{3r^2}(\cot A - \cot B).$$

11. By continuing the approximation in Art. 106 so as to include the terms involving r^4, shew that approximately

$$\cos A = \cos A' - \frac{\beta\gamma\sin^2 A'}{6r^2} + \frac{\beta\gamma(\alpha^2 - 3\beta^2 - 3\gamma^2)\sin^2 A'}{180r^4}.$$

12. From the preceding result shew that if $A = A' + \theta$ then approximately

$$\theta = \frac{\beta\gamma\sin A'}{6r^2}\left(1 + \frac{7\beta^2 + 7\gamma^2 + \alpha^2}{120r^2}\right).$$

X

GEODETICAL OPERATIONS.

114. One of the most important applications of Trigonometry, both Plane and Spherical, is to the determination of the figure and dimensions of the Earth itself, and of any portion of its surface. We shall give a brief outline of the subject, and for further information refer to Woodhouse's *Trigonometry*, to the article *Geodesy* in the *English Cyclopædia*, and to Airy's treatise on the *Figure of the Earth* in the *Encyclopædia Metropolitana*. For practical knowledge of the details of the operations it will be necessary to study some of the published accounts of the great surveys which have been effected in different parts of the world, as for example, the *Account of the measurement of two sections of the Meridional arc of India*, by Lieut.-Colonel Everest, 1847; or the *Account of the Observations and Calculations of the Principal Triangulation in the Ordnance Survey of Great Britain and Ireland*, 1858.

115. An important part of any survey consists in the measurement of a horizontal line, which is called a *base*. A level plain of a few miles in length is selected and a line is measured on it with every precaution to ensure accuracy. Rods of deal, and of metal, hollow tubes of glass, and steel chains, have been used in different surveys; the temperature is carefully observed during the operations, and allowance is made for the varying lengths of the rods or chains, which arise from variations in the temperature.

116. At various points of the country suitable stations are selected and signals erected; then by supposing lines to be drawn connecting the signals, the country is divided into a series of triangles. The angles of these triangles are observed, that is, the angles which any two signals subtend at a third. For example, suppose A and B to denote the extremities of the *base*, and C a signal at a third point visible from A and B; then in the triangle ABC the angles ABC and BAC are observed, and then AC and BC can be calculated. Again, let D be a signal at a fourth point, such that it is visible from C and A; then the angles ACD and CAD are observed, and as AC is known, CD and AD can be calculated.

117. Besides the original *base* other lines are measured in convenient parts of the country surveyed, and their measured lengths are compared with their lengths obtained by calculation through a series of triangles from the original base. The degree of closeness with which the measured length agrees with the calculated length is a test of the accuracy of the survey. During the progress of the Ordnance Survey of Great Britain and Ireland, several lines have been measured; the last two are, one near Lough Foyle in Ireland, which was measured in 1827 and 1828, and one on Salisbury Plain, which was measured in 1849. The line near Lough Foyle is nearly 8 miles long, and the line on Salisbury Plain is nearly 7 miles long; and the difference between the length of the line on Salisbury Plain as measured and as calculated from the Lough Foyle base is less than 5 inches (*An Account of the Observations* ... page 419).

118. There are different methods of effecting the calculations for determining the lengths of the sides of all the triangles in the survey. One method is to use the exact formulæ of Spherical Trigonometry. The radius of the Earth may be considered known very approximately; let this radius be denoted by r, then if α be the length of any arc the circular measure of the angle which the arc subtends at the centre of the earth is $\frac{\alpha}{r}$. The formulæ of Spherical Trigonometry gives expressions for the trigonometrical functions of $\frac{\alpha}{r}$, so that $\frac{\alpha}{r}$ may be found and then α. Since in practice $\frac{\alpha}{r}$ is always very small, it becomes necessary to pay attention to the methods of securing accuracy in calculations which involve the logarithmic trigonometrical functions of small angles (*Plane Trigonometry*, Art. 205).

Instead of the exact calculation of the triangles by Spherical Trigonometry, various methods of approximation have been proposed; only two of these methods however have been much used. One method of approximation consists in deducing from the angles of the spherical triangles the angles of the *chordal triangles*, and then computing the latter triangles by Plane Trigonometry (see Art. 105). The other method of approximation consists in the use of Legendre's Theorem (see Art. 106).

119. The three methods which we have indicated were all used by Delambre in calculating the triangles in the French survey (*Base du Système Métrique*, Tome III. page 7). In the earlier operations of the Trigonometrical survey of Great Britain and Ireland, the triangles were calculated by the chord method; but this has been for many years discontinued, and in place of it Legendre's Theorem has been universally adopted (*An Account of the Observations* ... page 244). The triangles in the Indian Survey are stated by Lieut.-Colonel Everest to be computed on Legendre's Theorem. (*An Account of the Measurement* ... page CLVIII.)

120. If the three angles of a plane triangle be observed, the fact that their sum ought to be equal to two right angles affords a test of the accuracy with which the observations are made. We shall proceed to shew how a test of the accuracy of observations of the angles of a spherical triangle formed on the Earth's surface may be obtained by means of the *spherical excess.*

121. *The area of a spherical triangle formed on the Earth's surface being known in square feet, it is required to establish a rule for computing the spherical excess in seconds.*

Let n be the number of seconds in the spherical excess, s the number of square feet in the area of the triangle, r the number of feet in the radius of the Earth. Then if E be the circular measure of the spherical excess,

$$s = Er^2,$$

and $$E = \frac{n\pi}{180 \,.\, 60 \,.\, 60} = \frac{n}{206265} \text{ approximately;}$$

therefore $$s = \frac{nr^2}{206265}.$$

Now by actual measurement the mean length of a degree on the Earth's surface is found to be 365155 feet; thus

$$\frac{\pi r}{180} = 365155.$$

With the value of r obtained from this equation it is found by logarithmic calculation, that

$$\log n = \log s - 9.326774.$$

Hence n is known when s is known.

This formula is called General Roy's rule, as it was used by him in the Trigonometrical survey of Great Britain and Ireland. Mr Davies, however, claims it for Mr Dalby. (See Hutton's *Course of Mathematics*, by Davies, Vol. II. p. 47.)

122. In order to apply General Roy's rule, we must know the area of the spherical triangle. Now the area is not known *exactly* unless the elements of the spherical triangle are known *exactly*; but it is found that in such cases as occur in practice an approximate value of the area is sufficient. Suppose, for example, that we use the area of the *plane triangle* considered in Legendre's Theorem, instead of the area of the *Spherical Triangle itself;* then it appears from Art. 109, that the error is approximately denoted by the fraction $\frac{\alpha^2 + \beta^2 + \gamma^2}{24r^2}$ of the former area, and this fraction is less than .0001, if the sides do not exceed 100 miles in length. Or again, suppose we want to estimate the influence of errors in the angles on the calculation of the area; let the circular measure of an error be h, so that instead of $\frac{\alpha\beta \sin C}{2}$ we ought to use $\frac{\alpha\beta \sin(C+h)}{2}$; the error then bears to the area approximately the ratio expressed by $h \cot C$. Now in modern observations h will not exceed the circular measure of a few seconds, so that, if C be not very small, $h \cot C$ is practically insensible.

123. The following example was selected by Woodhouse from the triangles of the English survey, and has been adopted by other writers. The observed angles of a triangle being respectively $42^\circ\ 2'\ 32''$, $67^\circ\ 55'\ 39''$, $70^\circ\ 1'\ 48''$, the sum of the errors made in the observations is required, supposing the side opposite to the angle A to be 27404.2 feet. The area is calculated from the expression

$\dfrac{a^2 \sin B \sin C}{2 \sin A}$, and by General Roy's rule it is found that $n = .23$. Now the sum of the observed angles is $180° - 1''$, and as it ought to have been $180° + .23''$, it follows that the sum of the errors of the observations is $1''.23$. This total error may be distributed among the observed angles in such proportion as the opinion of the observer may suggest; one way is to increase each of the observed angles by one-third of $1''.23$, and take the angles thus corrected for the true angles.

124. An investigation has been made with respect to the form of a triangle, in which errors in the observations of the angles will exercise the least influence on the lengths of the sides, and although the reasoning is allowed to be vague it may be deserving of the attention of the student. Suppose the three angles of a triangle observed, and one side, as a, known, it is required to find the form of the triangle in order that the other sides may be least affected by errors in the observations. The spherical excess of the triangle may be supposed known with sufficient accuracy for practice, and if the sum of the observed angles does not exceed two right angles by the proper spherical excess, let these angles be altered by adding the same quantity to each, so as to make their sum correct. Let A, B, C be the angles thus furnished by observation and altered if necessary; and let δA, δB and δC denote the respective errors of A, B and C. Then $\delta A + \delta B + \delta C = 0$, because by supposition the sum of A, B and C is correct. Considering the triangle as approximately plane, the true value of the side c is $\dfrac{a \sin(C + \delta C)}{\sin(A + \delta A)}$, that is, $\dfrac{a \sin(C + \delta C)}{\sin(A - \delta B - \delta C)}$. Now approximately

$$\sin(C + \delta C) = \sin C + \delta C \cos C, \quad (\textit{Plane Trig.}\ \text{Chap. XII.}),$$
$$\sin(A - \delta B - \delta C) = \sin A - (\delta B + \delta C) \cos A.$$

Hence approximately

$$c = \frac{a \sin C}{\sin A}\left\{1 + \delta C \cot C\right\}\left\{1 - (\delta B + \delta C) \cot A\right\}^{-1}$$
$$= \frac{a \sin C}{\sin A}\left\{1 + \delta B \cot A + \delta C(\cot C + \cot A)\right\};$$

and $\cot C + \cot A = \dfrac{\sin(A + C)}{\sin A \sin C} = \dfrac{\sin B}{\sin A \sin C}$ approximately.

Hence the error of c is approximately

$$\frac{a\sin B}{\sin^2 A}\delta C + \frac{a\sin C\cos A}{\sin^2 A)}\delta B.$$

Similarly the error of b is approximately

$$\frac{a\sin C}{\sin^2 A}\delta B + \frac{a\sin B\cos A}{\sin^2 A}\delta C.$$

Now it is impossible to assign exactly the signs and magnitudes of the errors δB and δC, so that the reasoning must be vague. It is obvious that to make the error small $\sin A$ must not be small. And as the sum of δA, δB and δC is zero, two of them must have the same sign, and the third the opposite sign; we may therefore consider that it is more probable than any two as δB and δC have different signs, than that they have the same sign.

If δB and δC have different signs the errors of b and c will be less when $\cos A$ is positive than when $\cos A$ is negative; A therefore ought to be less than a right angle. And if δB and δC are probably not very different, B and C should be nearly equal. These conditions will be satisfied by a triangle differing not much from an equilateral triangle.

If two angles only, A and B, be observed, we obtain the same expressions as before for the errors in b and c; but we have no reason for considering that δB and δC are of different signs rather than of the same sign. In this case then the supposition that A is a right angle will probably make the errors smallest.

125. The preceding article is taken from the Treatise on Trigonometry in the *Encyclopædia Metropolitana.* The least satisfactory part is that in which it is considered that δB and δC may be supposed nearly equal; for since $\delta A + \delta B + \delta C = 0$, if we suppose δB and δC nearly equal and of opposite signs, we do in effect suppose $\delta A = 0$ nearly; thus in observing three angles, we suppose that in one observation a certain error is made, in a second observation the same numerical error is made but with an opposite sign, and in the remaining observation no error is made.

126. We have hitherto proceeded on the supposition that the Earth is a sphere; it is however approximately a spheroid of small eccentricity. For the small corrections which must in consequence be introduced into the calculations we must refer to the works named in Art. 114. One of the results obtained is that the error caused by regarding the Earth as a sphere instead of a spheroid increases with the departure of the triangle from the well-conditioned or equilateral form (*An Account of the Observations* ... page 243). Under certain circumstances the spherical excess is the same on a spheroid as on a sphere (*Figure of the Earth* in the *Encyclopædia Metropolitana*, pages 198 and 215).

127. In geodetical operations it is sometimes required to determine the horizontal angle between two points, which are at a small angular distance from the horizon, the angle which the objects subtend being known, and also the angles of elevation or depression.

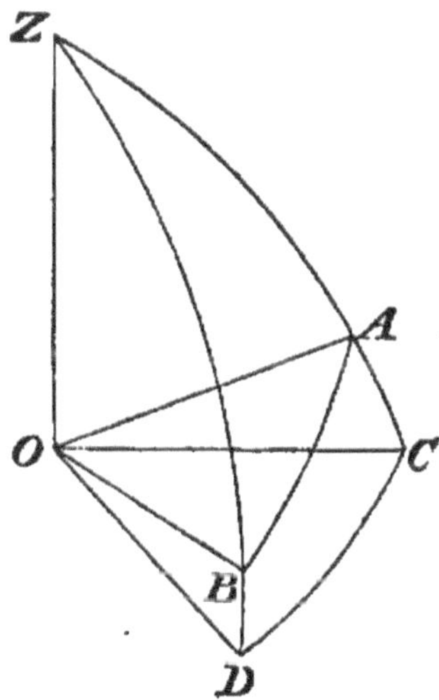

Suppose OA and OB the directions in which the two points are seen from O; and let the angle AOB be observed. Let OZ be the direction at right angles to the observer's horizon; describe a sphere round O as a centre, and let vertical planes through OA and OB meet the horizon at OC and OD respectively: then the angle COD is required.

Let $AOB = \theta$, $COD = \theta + x$, $AOC = h$, $BOD = k$; from the triangle AZB

$$\cos AZB = \frac{\cos\theta - \cos ZA \cos ZB}{\sin ZA \sin ZB} = \frac{\cos\theta - \sin h \sin k}{\cos h \cos k};$$

and $\cos AZB = \cos COD = \cos(\theta + x)$; thus

$$\cos(\theta + x) = \frac{\cos\theta - \sin h \, \sin k}{\cos h \, \cos k}.$$

This formula is exact; by approximation we obtain

$$\cos\theta - x\sin\theta = \frac{\cos\theta - hk}{1 - \frac{1}{2}(h^2 + k^2)};$$

therefore $$x\sin\theta = hk - \tfrac{1}{2}(h^2 + k^2)\cos\theta, \text{ nearly},$$

and $$x = \frac{2hk - (h^2 + k^2)(cos^2\frac{1}{2}\theta - \sin^2\frac{1}{2}\theta)}{2\sin\theta}$$

$$= \tfrac{1}{4}(h + k)^2 \tan\tfrac{1}{2}\theta - \tfrac{1}{4}(h - k)^2 \cot\tfrac{1}{2}\theta.$$

This process, by which we find the angle COD from the angle AOB, is called *reducing an angle to the horizon.*

XI

ON SMALL VARIATIONS IN THE PARTS OF A SPHERICAL TRIANGLE.

128. It is sometimes important to know what amount of error will be introduced into one of the calculated parts of a triangle by reason of any small error which may exist in the given parts. We will here consider an example.

129. *A side and the opposite angle of a spherical triangle remain constant: determine the connexion between the small variations of any other pair of elements.*

Suppose C and c to remain constant.

(1) Required the connexion between the small variations of the other sides. We suppose a and b to denote the sides of one triangle which can be formed with C and c as fixed elements, and $a + \delta a$ and $b + \delta b$ to denote the sides of another such triangle; then we require the ratio of δa to δb when both are extremely small. We have

$$\cos c = \cos a \cos b + \sin a \sin b \cos C,$$

and $$\cos c = \cos(a + \delta a) \cos(b + \delta b) + \sin(a + \delta a) \sin(b + \delta b) \cos C;$$

also $$\cos(a + \delta a) = \cos a - \sin a\, \delta a, \text{ nearly},$$

and $$\sin(a+\delta a) = \sin a + \cos a\,\delta a, \text{ nearly},$$

with similar formulæ for $\cos(b+\delta b)$ and $\sin(b+\delta b)$. (See *Plane Trigonometry*, Chap. XII.) Thus

$$\cos c = (\cos a - \sin a\,\delta a)(\cos b - \sin b\,\delta b) \\ + (\sin a + \cos a\,\delta a)(\sin b + \cos b\,\delta b)\cos C.$$

Hence by subtraction, if we neglect the product δa, δb,

$$0 = \delta a(\sin a\,\cos b - \cos a\,\sin b\,\cos C) \\ + \delta b(\sin b\,\cos a - \cos b\,\sin a\,\cos C);$$

this gives the ratio of δa to δb in terms of a, b, C. We may express the ratio more simply in terms of A and B; for, dividing by $\sin a \sin b$, we get from Art. 44,

$$\frac{\delta a}{\sin a}\cot B\,\sin C + \frac{\delta b}{\sin b}\cot A\,\sin C = 0;$$

therefore $$\delta a\,\cos B + \delta b\,\cos A = 0.$$

(2) Required the connexion between the small variations of the other angles. In this case we may by means of the polar triangle deduce from the result just found, that

$$\delta A\,\cos b + \delta B\,\cos a = 0;$$

this may also be found independently as before.

(3) Required the connexion between the small variations of a side and the opposite angle $(A,\ a)$.

Here $$\sin A \sin c = \sin C\,\sin a,$$

and $$\sin(A+\delta A)\sin c = \sin C\,\sin(a+\delta a);$$

hence by subtraction

$$\cos A\,\sin c\,\delta A = \sin C\,\cos a\,\delta a,$$

and therefore $$\delta A \cot A = \delta a \cot a.$$

(4) Required the connexion between the small variations of a side and the adjacent angle (a, B).

We have $$\cot C \sin B = \cot c \sin a - \cos B \cos a;$$

proceeding as before we obtain

$$\cot C \cos B \delta B = \cot c \cos a \delta a + \cos B \sin a \delta a + \cos a \sin B \delta B;$$

therefore

$$(\cot C \cos B - \cos a \sin B)\delta B = (\cot c \cos a + \cos B \sin a)\delta a;$$

therefore $$-\frac{\cos A}{\sin C}\delta B = \frac{\cos b}{\sin c}\delta a;$$

therefore $$\delta B \cos A = -\delta a \cot b \sin B.$$

130. Some more examples are proposed for solution at the end of this Chapter; as they involve no difficulty they are left for the exercise of the student.

EXAMPLES.

1. In a spherical triangle, if C and c remain constant while a and b receive the small increments δa and δb respectively, shew that

$$\frac{\delta a}{\sqrt{(1 - n^2 \sin^2 a)}} + \frac{\delta b}{\sqrt{(1 - n^2 \sin^2 b)}} = 0 \text{ where } n = \frac{\sin C}{\sin c}.$$

2. If C and c remain constant, and a small change be made in a, find the consequent changes in the other parts of the triangle. Find also the change in the area.

3. Supposing A and c to remain constant, prove the following equations, connecting the small variations of pairs of the other elements:

$$\sin C \delta b = \sin a \delta B, \quad \delta b \sin C = -\delta C \tan a, \quad \delta a \tan C = \delta B \sin a,$$
$$\delta a \tan C = -\delta C \tan a, \quad \delta b \cos C = \delta a, \quad \delta B \cos a = -\delta C.$$

4. Supposing b and c to remain constant, prove the following equations

connecting the small variations of pairs of the other elements:

$$\delta B \tan C = \delta C \tan B, \qquad \delta a \cot C = -\delta B \sin a,$$

$$\delta a = \delta A \sin c \sin B, \qquad \delta A \sin B \cos C = -\delta B \sin A.$$

5. Supposing B and C to remain constant, prove the following equations connecting the small variations of pairs of the other elements:

$$\delta b \tan c = \delta c \tan b, \qquad \delta A \cot c = \delta b \sin A,$$

$$\delta A = \delta a \sin b \sin C, \qquad \delta a \sin B \cos c = \delta b \sin A.$$

6. If A and C are constant, and b be increased by a small quantity, shew that a will be increased or diminished according as c is less or greater than a quadrant.

XII

ON THE CONNEXION OF FORMULÆ IN PLANE AND SPHERICAL TRIGONOMETRY.

131. The student must have perceived that many of the results obtained in *Spherical* Trigonometry resemble others with which he is familiar in *Plane* Trigonometry. We shall now pay some attention to this resemblance. We shall first shew how we may deduce formulæ in Plane Trigonometry from formulæ in Spherical Trigonometry; and we shall then investigate some theorems in Spherical Trigonometry which are interesting principally on account of their connexion with known results in Plane Geometry and Trigonometry.

132. *From any formula in Spherical Trigonometry involving the elements of a triangle, one of them being a side, it is required to deduce the corresponding formula in Plane Trigonometry.*

Let α, β, γ be the lengths of the sides of the triangle, r the radius of the sphere, so that $\frac{\alpha}{r}$, $\frac{\beta}{r}$, $\frac{\gamma}{r}$ are the circular measures of the sides of the triangle; expand the functions of $\frac{\alpha}{r}$, $\frac{\beta}{r}$, $\frac{\gamma}{r}$ which occur in any proposed formula in powers of $\frac{\alpha}{r}, \frac{\beta}{r}, \frac{\gamma}{r}$ respectively; then if we suppose r to become indefinitely great, the limiting form of the proposed formula will be a relation in Plane Trigonometry.

For example, in Art. 106, from the formula

$$\cos A = \frac{\cos a - \cos b \cos c}{\sin b \sin c}$$

we deduce

$$\cos A = \frac{\beta^2 + \gamma^2 - \alpha^2}{2\beta\gamma} + \frac{\alpha^4 + \beta^4 + \gamma^4 - 2\alpha^2\beta^2 - 2\beta^2\gamma^2 - 2\gamma^2\alpha^2}{24\beta\gamma r^2} + \ldots;$$

now suppose r to become infinite; then ultimately

$$\cos A = \frac{\beta^2 + \gamma^2 - \alpha^2}{2\beta\gamma};$$

and this is the expression for the cosine of the angle of a plane triangle in terms of the sides.

Again, in Art. 110, from the formula

$$\frac{\sin A}{\sin B} = \frac{\sin a}{\sin b}$$

we deduce

$$\frac{\sin A}{\sin B} = \frac{\alpha}{\beta} + \frac{\alpha(\beta^2 - \alpha^2)}{6\beta r^2} + \ldots;$$

now suppose r to become infinite; then ultimately

$$\frac{\sin A}{\sin B} = \frac{\alpha}{\beta},$$

that is, in a plane triangle the sides are as the sines of the opposite angles.

133. *To find the equation to a small circle of the sphere.*

The student can easily draw the required diagram.

Let O be the pole of a small circle, S a fixed point on the sphere, SX a fixed great circle of the sphere. Let $OS = \alpha$, $OSX = \beta$; then the position of O is determined by means of these angular co-ordinates α and β. Let P be any point on the circumference of the small circle, $PS = \theta$, $PSX = \phi$, so that θ and

ϕ are the angular co-ordinates of P. Let $OP = r$. Then from the triangle OSP

$$\cos r = \cos\alpha\cos\theta + \sin\alpha\sin\theta\cos(\phi - \beta); \tag{1}$$

this gives a relation between the angular co-ordinates of any point on the circumference of the circle.

If the circle be a great circle then $r = \frac{\pi}{2}$; thus the equation becomes

$$0 = \cos\alpha\cos\theta + \sin\alpha\sin\theta\cos(\phi - \beta). \tag{2}$$

It will be observed that the angular co-ordinates here used are analogous to the *latitude* and *longitude* which serve to determine the positions of places on the Earth's surface; θ is the *complement of the latitude* and ϕ is the *longitude.*

134. Equation (1) of the preceding Article may be written thus:

$$\cos r\left(\cos^2\frac{\theta}{2} + \sin^2\frac{\theta}{2}\right)$$
$$= \cos\alpha\left(\cos^2\frac{\theta}{2} - \sin^2\frac{\theta}{2}\right) + 2\sin\alpha\sin\frac{\theta}{2}\cos\frac{\theta}{2}\cos(\phi - \beta).$$

Divide by $\cos^2\frac{\theta}{2}$ and rearrange; hence

$$\tan^2\frac{\theta}{2}(\cos r + \cos\alpha) - 2\tan\frac{\theta}{2}\sin\alpha\cos(\phi - \beta) + \cos r - \cos\alpha = 0.$$

Let $\tan\frac{\theta_1}{2}$ and $\tan\frac{\theta_2}{2}$ denote the values of $\tan\frac{\theta}{2}$ found from this quadratic equation; then by *Algebra*, Chapter XXII.

$$\tan\frac{\theta_1}{2}\tan\frac{\theta_2}{2} = \frac{\cos r - \cos\alpha}{\cos r + \cos\alpha} = \tan\frac{\alpha + r}{2}\tan\frac{\alpha - r}{2}.$$

Thus the value of the product $\tan\frac{\theta_1}{2}\tan\frac{\theta_2}{2}$ is *independent* of ϕ; this result corresponds to the well-known property of a circle in Plane Geometry which is demonstrated in Euclid III. 36 *Corollary.*

135. Let three arcs OA, OB, OC meet at a point. From any point P in OB draw PM perpendicular to OA, and PN perpendicular to OC. The student can easily draw the required diagram.

Then, by Art. 65,

$$\sin PM = \sin OP \sin AOB, \quad \sin PN = \sin OP \sin COB;$$

therefore
$$\frac{\sin PM}{\sin PN} = \frac{\sin AOB}{\sin COB}.$$

Thus the ratio of $\sin PM$ to $\sin PN$ is independent of the position of P on the arc OB.

136. Conversely suppose that from any other point p arcs pm and pn are drawn perpendicular to OA and OC respectively; then if

$$\frac{\sin pm}{\sin pn} = \frac{\sin PM}{\sin PN},$$

it will follow that p is on the same great circle as O and P.

137. From two points P_1 and P_2 arcs are drawn perpendicular to a fixed arc; and from a point P on the same great circle as P_1 and P_2 a perpendicular is drawn to the same fixed arc. Let $PP_1 = \theta_1$ and $PP_2 = \theta_2$; and let the perpendiculars drawn from P, P_1, and P_2 be denoted by x, x_1 and x_2. Then will

$$\sin x = \frac{\sin\theta_2}{\sin(\theta_1+\theta_2)}\sin x_1 + \frac{\sin\theta_1}{\sin(\theta_1+\theta_2)}\sin x_2.$$

Let the arc P_1P_2, produced if necessary, cut the fixed arc at a point O; let α denote the angle between the arcs. We will suppose that P_1 is between O and P_2, and that P is between P_1 and P_2.

Then, by Art. 65,

$$\begin{aligned}\sin x_1 = \sin\alpha\sin OP_1 &= \sin\alpha\sin(OP-\theta_1)\\ &= \sin\alpha(\sin OP\cos\theta_1 - \cos OP\sin\theta_1);\\ \sin x_2 = \sin\alpha\sin OP_2 &= \sin\alpha\sin(OP+\theta_2)\\ &= \sin\alpha(\sin OP\cos\theta_2 + \cos OP\sin\theta_2).\end{aligned}$$

Multiply the former by $\sin\theta_2$, and the latter by $\sin\theta_1$, and add; thus

$$\begin{aligned}\sin\theta_2 \sin x_1 + \sin\theta_1 \sin x_2 &= \sin(\theta_1+\theta_2)\sin\alpha \sin OP \\ &= \sin(\theta_1+\theta_2)\sin x.\end{aligned}$$

The student should convince himself by examination that the result holds for all relative positions of P, P_1 and P_2, when due regard is paid to algebraical signs.

138. The principal use of Art. 137 is to determine whether three given points are on the same great circle; an illustration will be given in Art. 146.

139. *The arcs drawn from the angles of a spherical triangle perpendicular to the opposite sides respectively meet at a point.*

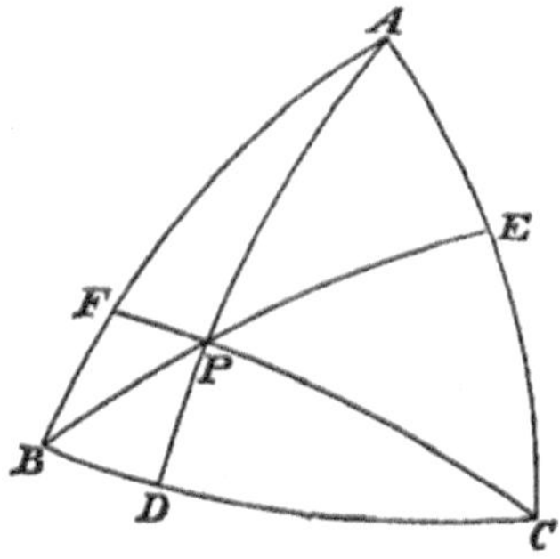

Let CF be perpendicular to AB. From F suppose arcs drawn perpendicular to CB and CA respectively; denote the former by ξ and the latter by η. Then, by Art. 135,

$$\frac{\sin\xi}{\sin\eta} = \frac{\sin FCB}{\sin FCA}.$$

But, by Art. 65,

$$\cos B = \cos CF \sin FCB, \quad \cos A = \cos CF \sin FCA;$$

therefore $$\frac{\sin\xi}{\sin\eta} = \frac{\cos B}{\cos A} = \frac{\cos B \cos C}{\cos A \cos C}.$$

And if from *any* point in CF arcs are drawn perpendicular to CB and CA

respectively, the ratio of the sine of the former perpendicular to the sine of the latter perpendicular is equal to $\frac{\sin\xi}{\sin\eta}$ by Art. 135.

In like manner suppose AD perpendicular to BC; then if from any point in AD arcs are drawn perpendicular to AC and AB respectively, the ratio of the sine of the former perpendicular to the sine of the latter perpendicular is equal to $\frac{\cos A\cos C}{\cos A\cos B}$.

Let CF and AD meet at P, and from P let perpendiculars be drawn on the sides a, b, c of the triangle; and denote these perpendiculars by x, y, z respectively: then we have shewn that

$$\frac{\sin x}{\sin y} = \frac{\cos B\cos C}{\cos A\cos C},$$

and that

$$\frac{\sin y}{\sin z} = \frac{\cos A\cos C}{\cos A\cos B};$$

hence it follows that

$$\frac{\sin x}{\sin z} = \frac{\cos B\cos C}{\cos B\cos A},$$

and this shews that the point P is on the arc drawn from B perpendicular to AC.

Thus the three perpendiculars meet at a point, and this point is determined by the relations

$$\frac{\sin x}{\cos B\cos C} = \frac{\sin y}{\cos C\cos A} = \frac{\sin z}{\cos A\cos B}.$$

140. In the same manner it may be shewn that the arcs drawn from the angles of a spherical triangle to the middle points of the opposite sides meet at a point; and if from this point arcs x, y, z are drawn perpendicular to the sides a, b, c respectively,

$$\frac{\sin x}{\sin B\sin C} = \frac{\sin y}{\sin C\sin A} = \frac{\sin z}{\sin A\sin B}.$$

141. It is known in Plane Geometry that a certain circle touches the inscribed and escribed circles of any triangle; this circle is called the *Nine points circle*: see *Appendix to Euclid*, pages 317, 318, and *Plane Trigonometry*, Chapter XXIV.

We shall now shew that a small circle can always be determined on the sphere to touch the inscribed and escribed circles of any spherical triangle.

142. Let α denote the distance from A of the pole of the small circle inscribed within a spherical triangle ABC. Suppose that a small circle of angular radius ρ touches this inscribed circle internally; let β be the distance from A of the pole of this touching circle; let γ be the angle between arcs drawn from A to the pole of the inscribed circle and the pole of the touching circle respectively. Then we must have

$$\cos(\rho - r) = \cos\alpha\cos\beta + \sin\alpha\sin\beta\cos\gamma. \tag{1}$$

Suppose that this touching circle also touches externally the escribed circle of angular radius r_1; then if α_1 denote the distance from A of the pole of this escribed circle, we must have

$$\cos(\rho + r_1) = \cos\alpha_1\cos\beta + \sin\alpha_1\sin\beta\cos\gamma. \tag{2}$$

Similarly, if α_2 and α_3 denote the distances from A of the poles of the other escribed circles, in order that the touching circle may touch these escribed circles externally, we must also have

$$\cos(\rho + r_2) = \cos\alpha_2\cos\beta + \sin\alpha_2\sin\beta\cos\left(\frac{\pi}{2} - \gamma\right), \tag{3}$$

$$\cos(\rho + r_3) = \cos\alpha_3\cos\beta + \sin\alpha_3\sin\beta\cos\left(\frac{\pi}{2} + \gamma\right). \tag{4}$$

We shall shew that real values of ρ, β, and γ can be found to satisfy these four equations.

Eliminate $\cos\gamma$ from (1) and (2); thus

$$\begin{aligned}\cos\rho(\cos r\sin\alpha_1 - \cos r_1\sin\alpha) + \sin\rho(\sin r\sin\alpha_1 + \sin r_1\sin\alpha)\\ = \cos\beta(\cos\alpha\sin\alpha_1 - \cos\alpha_1\sin\alpha).\end{aligned} \tag{5}$$

Suppose that the inscribed circle touches AB at the distance m from A, and that the escribed circle of angular radius r_1 touches AB at the distance m_1 from

A. Then, by Art. 65,

$$\cot\alpha = \cot m \cos\frac{A}{2}, \quad \cos\alpha = \cos r\cos m, \quad \sin r = \sin\alpha\sin\frac{A}{2};$$

therefore
$$\frac{\cos r}{\sin\alpha} = \frac{\cot\alpha}{\cos m} = \frac{1}{\sin m}\cos\frac{A}{2}.$$

Similarly we may connect α_1 and r_1 with m_1. Thus we obtain from (5)

$$\cos\rho\cos\frac{A}{2}\left(\frac{1}{\sin m} - \frac{1}{\sin m_1}\right) + 2\sin\rho\sin\frac{A}{2} = \cos\beta\cos\frac{A}{2}(\cot m - \cot m_1);$$

therefore
$$\cos\rho(\sin m_1 - \sin m) + 2\sin\rho\sin m\sin m_1\tan\frac{A}{2} = \cos\beta\sin(m_1 - m).$$

But by Arts. 89 and 90 we have $m = s - a$, and $m_1 = s$; therefore by the aid of Art. 45 we obtain

$$2\cos\rho\sin\frac{a}{2}\cos\frac{b+c}{2} + 2n\sin\rho = \cos\beta\sin a, \tag{6}$$

where n has the meaning assigned in Art. 46.

In like manner if we eliminate $\sin\gamma$ between (3) and (4), putting m_2 for $s-c$, and m_3 for $s-b$, we obtain

$$\cos\rho(\sin m_2 + \sin m_3) - 2\sin\rho\sin m_2\sin m_3\cot\frac{A}{2} = \cos\beta\sin(m_2 + m_3),$$

therefore
$$2\cos\rho\sin\frac{a}{2}\cos\frac{b-c}{2} - 2n\sin\rho = \cos\beta\sin a. \tag{7}$$

From (6) and (7) we get

$$\tan\rho = \frac{\sin\frac{a}{2}\sin\frac{b}{2}\sin\frac{c}{2}}{n} = \frac{1}{2}\tan R, \text{ by Art. 92} \tag{8}$$

and
$$\cos\beta = \frac{\cos\frac{b}{2}\cos\frac{c}{2}\cos\rho}{\cos\frac{a}{2}}. \tag{9}$$

We may suppose that $\cos\frac{a}{2}$ is not less than $\cos\frac{b}{2}$ or $\cos\frac{c}{2}$, so that we are sure of a possible value of $\cos\beta$ from (9).

It remains to shew that when ρ and β are thus determined, all the four fundamental equations are satisfied.

It will be observed that, ρ and β being considered known, $\cos\gamma$ can be found from (1) or (2), and $\sin\gamma$ can be found from (3) or (4): we must therefore shew that (1) and (2) give the *same* value for $\cos\gamma$, and that (3) and (4) give the *same* value for $\sin\gamma$; and we must also shew that these values satisfy the condition $\cos^2\gamma + \sin^2\gamma = 1$.

From (1) we have

$$\frac{\cos\rho\sin r}{\sin\alpha}\left(\cot r + \tan\rho - \cos m\cot r\frac{\cos\beta}{\cos\rho}\right) = \sin\beta\cos\gamma,$$

that is,

$$\frac{\cos\rho\sin\frac{A}{2}}{n}\left\{\sin s + \sin\tfrac{1}{2}a\sin\tfrac{1}{2}b\sin\tfrac{1}{2}c - \frac{\cos(s-a)\sin s\cos\frac{b}{2}\cos\frac{c}{2}}{\cos\frac{a}{2}}\right\}$$
$$= \sin\beta\cos\gamma;$$

this reduces to

$$\frac{\cos\rho\sin\frac{A}{2}}{n}\left\{\cos\frac{a}{2}\sin\frac{b+c}{2} - \frac{\sin(b+c)\cos\frac{b}{2}\cos\frac{c}{2}}{2\cos\frac{a}{2}}\right\} = \sin\beta\cos\gamma:$$

and it will be found that (2) reduces to the same; so that (1) and (2) give the same value for $\cos\gamma$.

In like manner it will be found that (3) and (4) agree in reducing to

$$\frac{\cos\rho\cos\frac{A}{2}}{n}\left\{\cos\frac{a}{2}\sin\frac{c-b}{2} - \frac{\sin(c-b)\cos\frac{b}{2}\cos\frac{c}{2}}{2\cos\frac{a}{2}}\right\} = \sin\beta\cos\gamma.$$

It only remains to shew that the condition $\cos^2\gamma + \sin^2\gamma = 1$ is satisfied.

Put k for $\dfrac{\cos\beta}{\cos\rho}$, that is for $\dfrac{\cos\frac{b}{2}\cos\frac{c}{2}}{\cos\frac{a}{2}}$;

put X for $\cot r\{1 - k\cos(s-a)\}$, and Y for $\cot r_1\{1 - k\cos s\}$.

Then (1) and (2) may be written respectively thus:

$$(X\cos\rho + \sin\rho)\sin\frac{A}{2} = \sin\beta\cos\gamma, \qquad (10)$$

$$(Y\cos\rho - \sin\rho)\sin\frac{A}{2} = \sin\beta\cos\gamma. \qquad (11)$$

From (10) and (11) by addition

$$(X+Y)\sin\frac{A}{2}\cos\rho = 2\sin\beta\cos\gamma;$$

therefore $$4\sin^2\beta\cos^2\gamma = (X^2+Y^2+2XY)\sin^2\frac{A}{2}\cos^2\rho. \qquad (12)$$

But from (10) and (11) by subtraction

$$(X-Y)\cos\rho = -2\sin\rho;$$

therefore $$(X^2+Y^2)\cos^2\rho = 4\sin^2\rho + 2XY\cos^2\rho.$$

Substitute in (12) and we obtain

$$\sin^2\beta\cos^2\gamma = (\sin^2\rho + XY\cos^2\rho)\sin^2\frac{A}{2}. \qquad (13)$$

Again, put

X_1 for $\cot r_2\{1 - k\cos(s-c)\}$, and Y_1 for $\cot r_3\{1 - k\cos(s-b)\}$.

Then (3) and (4) may be written respectively thus:

$$(X_1\cos\rho - \sin\rho)\cos\frac{A}{2} = \sin\beta\sin\gamma, \qquad (14)$$

$$(Y_1\cos\rho - \sin\rho)\cos\frac{A}{2} = -\sin\beta\sin\gamma. \qquad (15)$$

From (14) and (15) by subtraction

$$(X_1 - Y_1)\cos\frac{A}{2}\cos\rho = 2\sin\beta\sin\gamma,$$

and from (14) and (15) by addition,

$$(X_1 + Y_1)\cos\rho = 2\sin\rho,$$

whence

$$\sin^2\beta\sin^2\gamma = (\sin^2\rho - X_1Y_1\cos^2\rho)\cos^2\frac{A}{2}. \tag{16}$$

Hence from (13) and (16) it follows that we have to establish the relation

$$\sin^2\beta = \sin^2\rho + \left(XY\sin^2\frac{A}{2} - X_1Y_1\cos^2\frac{A}{2}\right)\cos^2\rho.$$

But $\sin^2\beta = 1 - \cos^2\beta = \sin^2\rho + \cos^2\rho - k^2\cos^2\rho$, so that the relation reduces to

$$1 - k^2 = XY\sin^2\frac{A}{2} - X_1Y_1\cos^2\frac{A}{2}.$$

Now

$$XY\sin^2\frac{A}{2} = \frac{\cot r\cot r_1\{1 - k\cos s\}\{1 - k\cos(s-a)\}\sin(s-b)\sin(s-c)}{\sin b\sin c}$$

$$= \frac{\{1 - k\cos s\}\{1 - k\cos(s-a)\}}{\sin b\sin c}.$$

Similarly $X_1Y_1\cos^2\dfrac{A}{2} = \dfrac{\{1 - k\cos(s-b)\}\{1 - k\cos(s-c)\}}{\sin b\sin c}$.

Subtract the latter from the former; then we obtain

$$\frac{k}{\sin b\sin c}\{\cos(s-b) + \cos(s-c) - \cos s - \cos(s-a)\}$$

$$+ \frac{k^2}{\sin b\sin c}\{\cos s\cos(s-a) - \cos(s-b)\cos(s-c)\},$$

that is $$\frac{2k\cos\frac{a}{2}}{\sin b\sin c}\left\{\cos\frac{b-c}{2} - \cos\frac{b+c}{2}\right\}$$

$$+\frac{k^2}{\sin b\sin c}\left\{\cos\frac{b+c+a}{2}\cos\frac{b+c-a}{2}-\cos\frac{a+c-b}{2}\cos\frac{a+b-c}{2}\right\}$$

that is $\dfrac{4\sin\frac{b}{2}\sin\frac{c}{2}\cos\frac{b}{2}\cos\frac{c}{2}}{\sin b\sin c}+\dfrac{k^2}{\sin b\sin c}\left\{\sin^2\frac{c-b}{2}-\sin^2\frac{c+b}{2}\right\}$,

that is $1-k^2$; which was to be shewn.

143. Thus the existence of a circle which touches the inscribed and escribed circles of any spherical triangle has been established.

The distance of the pole of this touching circle from the angles B and C of the triangle will of course be determined by formulæ corresponding to (9); and thus it follows that

$$\frac{\cos\frac{a}{2}\cos\frac{c}{2}\cos\rho}{\cos\frac{b}{2}}\quad\text{and}\quad\frac{\cos\frac{a}{2}\cos\frac{b}{2}\cos\rho}{\cos\frac{c}{2}},$$

must both be less than unity.

144. Since the circle which has been determined touches the inscribed circle internally and touches the escribed circles externally, it is obvious that it must meet all the sides of the spherical triangle. We will now determine the position of the points of meeting.

Suppose the touching circle intersects the side AB at points distant λ and μ respectively from A.

Then by Art. 134 we have

$$\tan\frac{\lambda}{2}\tan\frac{\mu}{2}=\frac{\cos\rho-\cos\beta}{\cos\rho+\cos\beta}=\frac{\cos\frac{a}{2}-\cos\frac{b}{2}\cos\frac{c}{2}}{\cos\frac{a}{2}+\cos\frac{b}{2}\cos\frac{c}{2}}. \tag{1}$$

In the same way we must have by symmetry

$$\tan\frac{c-\lambda}{2}\tan\frac{c-\mu}{2}=\frac{\cos\frac{b}{2}-\cos\frac{a}{2}\cos\frac{c}{2}}{\cos\frac{b}{2}+\cos\frac{a}{2}\cos\frac{c}{2}}. \tag{2}$$

From (2), when we substitute the value of $\tan\frac{\lambda}{2}\tan\frac{\mu}{2}$ given by (1), we obtain

$$\tan\frac{\lambda}{2}+\tan\frac{\mu}{2}=\frac{\cos^2\frac{a}{2}-\cos^2\frac{b}{2}\cos^2\frac{c}{2}+\cos^2\frac{b}{2}\sin^2\frac{c}{2}}{\cos\frac{b}{2}\sin\frac{c}{2}\left(\cos\frac{a}{2}+\cos\frac{b}{2}\cos\frac{c}{2}\right)}$$

$$=\frac{\cos\frac{a}{2}-\cos\frac{b}{2}\cos\frac{c}{2}}{\cos\frac{b}{2}\sin\frac{c}{2}}+\frac{\cos\frac{b}{2}\sin\frac{c}{2}}{\cos\frac{a}{2}+\cos\frac{b}{2}\cos\frac{c}{2}}. \qquad (3)$$

From (1) and (3) we see that we may put

$$\tan\frac{\lambda}{2}=\frac{\cos\frac{a}{2}-\cos\frac{b}{2}\cos\frac{c}{2}}{\cos\frac{b}{2}\sin\frac{c}{2}}. \qquad (4)$$

$$\tan\frac{\mu}{2}=\frac{\cos\frac{b}{2}\sin\frac{c}{2}}{\cos\frac{a}{2}+\cos\frac{b}{2}\cos\frac{c}{2}}. \qquad (5)$$

Similar formulæ of course hold for the points of intersection of the touching circle with the other sides.

145. Let z denote the perpendicular from the pole of the touching circle on AB; then

$$\sin z=\sin\beta\sin\left(\frac{A}{2}+\gamma\right)$$

$$=\sin\beta\left(\sin\frac{A}{2}\cos\gamma+\cos\frac{A}{2}\sin\gamma\right).$$

But from (2) and (3) of Art. 142 we have

$$\sin\beta\cos\gamma=\frac{\cos\rho\sin\frac{A}{2}}{n}\left(Z-\sin\frac{a}{2}\sin\frac{b}{2}\sin\frac{c}{2}\right),$$

where $$Z = \sin(s-a) - \cos s \sin(s-a) \cos\frac{b}{2} \cos\frac{c}{2} \sec\frac{a}{2},$$

and $$\sin\beta \sin\gamma = \frac{\cos\rho \cos\frac{A}{2}}{n} \left(Z_1 - \sin\frac{a}{2} \sin\frac{b}{2} \sin\frac{c}{2}\right),$$

where $$Z_1 = \sin(s-b) - \cos(s-c) \sin(s-b) \cos\frac{b}{2} \cos\frac{c}{2} \sec\frac{a}{2}.$$

Therefore

$$\sin z = \frac{\cos\rho}{n} \left\{Z \sin^2\frac{A}{2} + Z_1 \cos^2\frac{A}{2} - \sin\frac{a}{2} \sin\frac{b}{2} \sin\frac{c}{2}\right\}.$$

Now $Z \sin^2\frac{A}{2}$

$$= \frac{\sin(s-a)\sin(s-b)\sin(s-c)}{\sin b \sin c} \left\{1 - \cos s \cos\frac{b}{2} \cos\frac{c}{2} \sec\frac{a}{2}\right\},$$

and $Z_1 \cos^2\frac{A}{2}$

$$= \frac{\sin s \sin(s-a)\sin(s-b)}{\sin b \sin c} \left\{1 - \cos(s-c) \cos\frac{b}{2} \cos\frac{c}{2} \sec\frac{a}{2}\right\}.$$

Therefore $$Z \sin^2\frac{A}{2} + Z_1 \cos^2\frac{A}{2}$$
is equal to the product of

$$\frac{\sin(s-a)\sin(s-b)}{\sin b \sin c}$$

into

$$\sin(s-c) + \sin s - \cos\frac{b}{2} \cos\frac{c}{2} \sec\frac{a}{2} \left\{\sin(s-c)\cos s + \cos(s-c)\sin s\right\}$$

$$= \frac{\sin(s-a)\sin(s-b)}{\sin b \sin c} \left\{2 \sin\frac{a+b}{2} \cos\frac{c}{2} - \cos\frac{b}{2} \cos\frac{c}{2} \sec\frac{a}{2} \sin(2s-c)\right\}$$

$$= \frac{\sin(s-a)\sin(s-b)}{2 \sin b \sin\frac{c}{2}} \left\{2 \sin\frac{a+b}{2} - \sin(a+b) \cos\frac{b}{2} \sec\frac{a}{2}\right\}$$

$$= \frac{\sin(s-a)\sin(s-b)\sin\dfrac{a+b}{2}}{\sin b \sin\dfrac{c}{2}} \left\{ 1 - \frac{\cos\dfrac{a+b}{2}\cos\dfrac{b}{2}}{\cos\dfrac{a}{2}} \right\}$$

$$= \frac{\sin(s-a)\sin(s-b)\sin^2\dfrac{a+b}{2}\sin\dfrac{b}{2}}{\sin b \sin\dfrac{c}{2}\cos\dfrac{a}{2}} = \frac{\sin(s-a)\sin(s-b)\sin^2\dfrac{a+b}{2}}{2\cos\dfrac{a}{2}\cos\dfrac{b}{2}\sin\dfrac{c}{2}}.$$

Therefore

$$\sin z = \frac{\cos\rho}{n}\sin\frac{a}{2}\sin\frac{b}{2}\sin\frac{c}{2}\left\{ \frac{2\sin^2\dfrac{a+b}{2}\sin(s-a)\sin(s-b)}{\sin^2\dfrac{c}{2}\sin a\sin b} - 1 \right\}$$

$$= \frac{\cos\rho}{n}\sin\frac{a}{2}\sin\frac{b}{2}\sin\frac{c}{2}\left\{ 2\cos^2\frac{A-B}{2} - 1 \right\};\ \text{by (2) of Art. 54.}$$

Thus

$$\sin z = \frac{\cos\rho}{n}\sin\frac{a}{2}\sin\frac{b}{2}\sin\frac{c}{2}\cos(A-B)$$

$$= \sin\rho\cos(A-B).$$

Similar expressions hold for the perpendiculars from the pole of the touching circle on the other sides of the spherical triangle.

146. Let P denote the point determined in Art. 139; G the point determined in Art. 140, and N the pole of the touching circle. We shall now shew that P, G, and N are on a great circle.

Let x, y, z denote the perpendiculars from N on the sides a, b, c respectively of the spherical triangle; let x_1, y_1, z_1 denote the perpendiculars from P; and x_2, y_2, z_2 the perpendiculars from G. Then by Arts. 145, 139, and 140 we have

$$\frac{\sin x}{\cos(B-C)} = \frac{\sin y}{\cos(C-A)} = \frac{\sin z}{\cos(A-B)},$$

$$\frac{\sin x_1}{\cos B\cos C} = \frac{\sin y_1}{\cos C\cos A} = \frac{\sin z_1}{\cos A\cos B},$$

$$\frac{\sin x_2}{\sin B\sin C} = \frac{\sin y_2}{\sin C\sin A} = \frac{\sin z_2}{\sin A\sin B}.$$

Hence it follows that

$$\sin x = t_1 \sin x_1 + t_2 \sin x_2,$$
$$\sin y = t_1 \sin y_1 + t_2 \sin y_2,$$
$$\sin z = t_1 \sin z_1 + t_2 \sin z_2,$$

where t_1 and t_2 are certain quantities the values of which are not required for our purpose.

Therefore by Art. 137 a certain point *in the same great circle* as P and G is at the perpendicular distances x, y, z from the sides a, b, c respectively of the spherical triangle: and hence this point must be the point N.

147. The resemblance of the results which have been obtained to those which are known respecting the Nine points circle in Plane Geometry will be easily seen.

The result $\tan\rho = \frac{1}{2}\tan R$ corresponds to the fact that the radius of the Nine points circle is half the radius of the circumscribing circle of the triangle.

From equation (4) of Art. 144 by supposing the radius of the sphere to become infinite we obtain $\lambda = \frac{b^2+c^2-a^2}{2c}$: this corresponds to the fact that the Nine points circle passes through the feet of the perpendiculars from the angles of a triangle on the opposite sides.

From equation (5) of Art. 144 by supposing the radius of the sphere to become infinite we obtain $\mu = \frac{c}{2}$: this corresponds to the fact that the Nine points circle passes through the middle points of the sides of a triangle.

From Art. 145 by supposing the radius of the sphere to become infinite we obtain $z = \frac{1}{2}R\cos(A-B)$: this is a known property of the Nine points circle.

In Plane Geometry the points which correspond to the P, G, and N of Art. 146 are on a straight line.

148. The results which have been demonstrated with respect to the circle which touches the inscribed and escribed circles of a spherical triangle are mainly due to Dr Hart and Dr Salmon. See the *Quarterly Journal of Mathematics*, Vol. VI. page 67.

EXAMPLES.

1. From the formula $\sin\frac{a}{2} = \sqrt{\left\{\frac{-\cos S\cos(S-A)}{\sin B\sin C}\right\}}$ deduce the expression for the area of a plane triangle, namely $\frac{a^2\sin B\sin C}{2\sin A}$, when the radius of the sphere is indefinitely increased.

2. Two triangles ABC, abc, spherical or plane, equal in all respects, differ slightly in position: shew that

$$\cos ABb\cos BCc\cos CAa + \cos ACc\cos CBb\cos BAa = 0.$$

3. Deduce formulæ in Plane Trigonometry from Napier's Analogies.

4. Deduce formulæ in Plane Trigonometry from Delambre's Analogies.

5. From the formula $\cos\frac{c}{2}\cos\frac{A+B}{2} = \sin\frac{C}{2}\cos\frac{a+b}{2}$ deduce the area of a plane triangle in terms of the sides and one of the angles.

6. What result is obtained from Example 7 to Chapter VI., by supposing the radius of the sphere infinite?

7. From the angle C of a spherical triangle a perpendicular is drawn to the arc which joins the middle points of the sides a and b: shew that this perpendicular makes an angle $S - B$ with the side a, and an angle $S - A$ with the side b.

8. From each angle of a spherical triangle a perpendicular is drawn to the arc which joins the middle points of the adjacent sides. Shew that these perpendiculars meet at a point; and that if x, y, z are the perpendiculars from this point on the sides a, b, c respectively,

$$\frac{\sin x}{\sin(S-B)\sin(S-C)} = \frac{\sin y}{\sin(S-C)\sin(S-A)} = \frac{\sin z}{\sin(S-A)\sin(S-B)}.$$

9. Through each angle of a spherical triangle an arc is drawn so as to make the same angle with one side which the perpendicular on the base makes with the other side. Shew that these arcs meet at a point; and that if x, y, z are the perpendiculars from this point on the sides a, b, c respectively,

$$\frac{\sin x}{\cos A} = \frac{\sin y}{\cos B} = \frac{\sin z}{\cos C}.$$

10. Shew that the points determined in Examples 8 and 9, and the point N of Art. 146 are on a great circle.

State the corresponding theorem in Plane Geometry.

11. If one angle of a spherical triangle remains constant while the adjacent sides are increased, shew that the area and the sum of the angles are increased.

12. If the arcs bisecting two angles of a spherical triangle and terminated at the opposite sides are equal, the bisected angles will be equal provided their sum be less than 180°.

[Let BOD and COE denote these two arcs which are given equal. If the angles B and C are not equal suppose B the greater. Then CD is greater than BE by Art. 58. And as the angle OBC is greater than the angle OCB, therefore OC is greater than OB; therefore OD is greater than OE. Hence the angle ODC is greater than the angle OEB, by Example 11. Then construct a spherical triangle BCF on the other side of BC, equal to CBE. Since the angle ODC is greater than the angle OEB, the angle FDC is greater than the angle DFC; therefore CD is less than CF, so that CD is less than BE. See the corresponding problem in Plane Geometry in the *Appendix to Euclid*, page 317.]

XIII

POLYHEDRONS.

149. A polyhedron is a solid bounded by any number of plane rectilineal figures which are called its faces. A polyhedron is said to be *regular* when its faces are similar and equal regular polygons, and its solid angles equal to one another.

150. *If* S *be the number of solid angles in any polyhedron,* F *the number of its faces,* E *the number of its edges, then* $\mathrm{S} + \mathrm{F} = \mathrm{E} + 2$.

Take any point within the polyhedron as centre, and describe a sphere of radius r, and draw straight lines from the centre to each of the angular points of the polyhedron; let the points at which these straight lines meet the surface of the sphere be joined by arcs of great circles, so that the surface of the sphere is divided into as many polygons as the polyhedron has faces.

Let s denote the sum of the angles of any one of these polygons, m the number of its sides; then the area of the polygon is $r^2\{s-(m-2)\pi\}$ by Art. 99. The sum of the areas of all the polygons is the surface of the sphere, that is, $4\pi r^2$. Hence since the number of the polygons is F, we obtain

$$4\pi = \sum s - \pi \sum m + 2F\pi.$$

Now $\sum s$ denotes the sum of all the angles of the polygons, and is therefore equal to $2\pi\times$ the number of solid angles, that is, to $2\pi S$; and $\sum m$ is equal to the number of all the sides of all the polygons, that is, to $2E$, since every edge gives rise to an arc which is common to two polygons. Therefore

$$4\pi = 2\pi S - 2\pi E + 2F\pi;$$

therefore

$$S + F = E + 2.$$

151. *There can be only five regular polyhedrons.*

Let m be the number of sides in each face of a regular polyhedron, n the number of plane angles in each solid angle; then the entire number of plane angles is expressed by mF, or by nS, or by $2E$; thus

$$mF = nS = 2E, \text{ and } S + F = E + 2;$$

from these equations we obtain

$$S = \frac{4m}{2(m+n) - mn}, \quad E = \frac{2mn}{2(m+n) - mn}, \quad F = \frac{4n}{2(m+n) - mn}.$$

These expressions must be positive integers, we must therefore have $2(m+n)$ greater than mn; therefore

$$\frac{1}{m} + \frac{1}{n} \text{ must be greater than } \frac{1}{2};$$

but n cannot be less than 3, so that $\frac{1}{n}$ cannot be greater than $\frac{1}{3}$, and therefore $\frac{1}{m}$ must be greater than $\frac{1}{6}$; and as m must be an integer and cannot be less than 3, the only admissible values of m are 3, 4, 5. It will be found on trial that the only values of m and n which satisfy all the necessary conditions are the following: each regular polyhedron derives its name from the number of its plane faces.

m	n	S	E	F	Name of regular Polyhedron.
3	3	4	6	4	Tetrahedron or regular Pyramid.
4	3	8	12	6	Hexahedron or Cube.
3	4	6	12	8	Octahedron.
5	3	20	30	12	Dodecahedron.
3	5	12	30	20	Icosahedron.

It will be seen that the demonstration establishes something more than the enunciation states; for it is not assumed that the faces are equilateral and equiangular and all equal. It is in fact demonstrated that, *there cannot be more than five solids each of which has all its faces with the same number of sides, and all its solid angles formed with the same number of plane angles.*

152. *The sum of all the plane angles which form the solid angles of any polyhedron is* $2(S-2)\pi$.

For if m denote the number of sides in any face of the polyhedron, the sum of the interior angles of that face is $(m-2)\pi$ by Euclid I. 32, Cor. 1. Hence the sum of all the interior angles of all the faces is $\sum(m-2)\pi$, that is $\sum m\pi - 2F\pi$, that is $2(E-F)\pi$, that is $2(S-2)\pi$.

153. *To find the inclination of two adjacent faces of a regular polyhedron.*

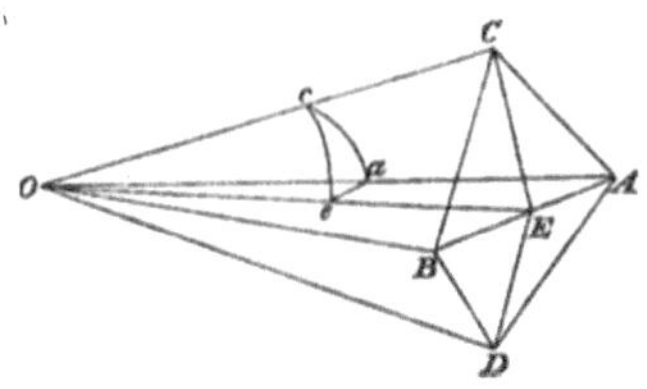

Let AB be the edge common to the two adjacent faces, C and D the centres of the faces; bisect AB at E, and join CE and DE; CE and DE will be perpendicular to AB, and the angle CED is the angle of inclination of the two adjacent faces; we shall denote it by I. In the plane containing CE and DE draw CO and DO at right angles to CE and DE respectively, and meeting at O; about O as centre describe a sphere meeting OA, OC, OE at a, c, e respectively, so that cae forms a spherical triangle. Since AB is perpendicular to CE and DE, it is perpendicular to the plane CED, therefore the plane AOB which contains AB is perpendicular to the plane CED; hence the angle cea of the spherical triangle is a right angle. Let m be the number of sides in each face of the polyhedron, n the number of the plane angles which form each solid angle. Then the angle $ace = ACE = \frac{2\pi}{2m} = \frac{\pi}{m}$; and the angle cae is half one of the n equal angles formed on the sphere round a, that is, $cae = \frac{2\pi}{2n} = \frac{\pi}{n}$. From the right-angled triangle cae

$$\cos cae = \cos cOe \sin ace,$$

that is
$$\cos\frac{\pi}{n} = \cos\left(\frac{\pi}{2}-\frac{I}{2}\right)\sin\frac{\pi}{m};$$

therefore
$$\sin\frac{I}{2} = \frac{\cos\frac{\pi}{n}}{\sin\frac{\pi}{m}}.$$

154. *To find the radii of the inscribed and circumscribed spheres of a regular polyhedron.*

Let the edge $AB = a$, let $OC = r$ and $OA = R$, so that r is the radius of the inscribed sphere, and R is the radius of the circumscribed sphere. Then

$$CE = AE\cot ACE = \frac{a}{2}\cot\frac{\pi}{m},$$

$$r = CE\tan CEO = CE\tan\frac{I}{2} = \frac{a}{2}\cot\frac{\pi}{m}\tan\frac{I}{2};$$

also
$$r = R\cos aOc = R\cot eca\cot eac = R\cot\frac{\pi}{m}\cot\frac{\pi}{n};$$

therefore
$$R = r\tan\frac{\pi}{m}\tan\frac{\pi}{n} = \frac{a}{2}\tan\frac{I}{2}\tan\frac{\pi}{n}.$$

155. *To find the surface and volume of a regular polyhedron.*

The area of one face of the polyhedron is $\frac{ma^2}{4}\cot\frac{\pi}{m}$, and therefore the surface of the polyhedron is $\frac{mFa^2}{4}\cot\frac{\pi}{m}$.

Also the volume of the pyramid which has one face of the polyhedron for base and O for vertex is $\frac{r}{3}\cdot\frac{ma^2}{4}\cot\frac{\pi}{m}$, and therefore the volume of the polyhedron is $\frac{mFra^2}{12}\cot\frac{\pi}{m}$.

156. *To find the volume of a parallelepiped in terms of its edges and their inclinations to one another.*

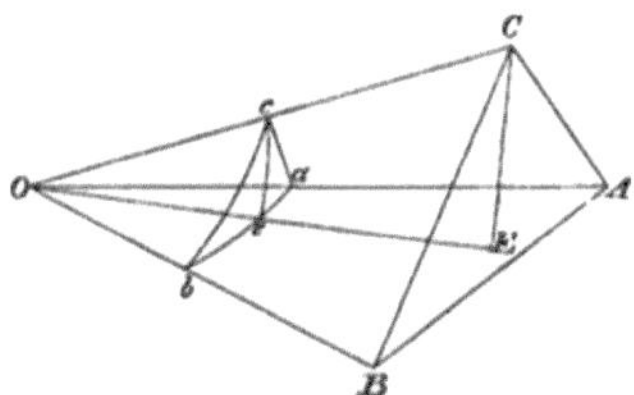

Let the edges be $OA = a$, $OB = b$, $OC = c$; let the inclinations be $BOC = a$, $COA = \beta$, $AOB = \gamma$. Draw CE perpendicular to the plane AOB meeting it at E. Describe a sphere with O as a centre, meeting OA, OB, OC, OE at a, b, c, e respectively.

The volume of the parallelepiped is equal to the product of its base and altitude $= ab\sin\gamma \,.\, CE = abc\sin\gamma\sin cOe$. The spherical triangle cae is right-angled at e; thus

$$\sin cOe = \sin cOa \sin cae = \sin\beta\sin cab,$$

and from the spherical triangle cab

$$\sin cab = \frac{\sqrt{(1-\cos^2\alpha-\cos^2\beta-\cos^2\gamma+2\cos\alpha\cos\beta\cos\gamma)}}{\sin\beta\sin\gamma};$$

therefore the volume of the parallelepiped

$$= abc\sqrt{(1-\cos^2\alpha-\cos^2\beta-\cos^2\gamma+2\cos\alpha\cos\beta\cos\gamma)}.$$

157. *To find the diagonal of a parallelepiped in terms of the three edges which it meets and their inclinations to one another.*

Let the edges be $OA = a$, $OB = b$, $OC = c$; let the inclinations be $BOC = \alpha$, $COA = \beta$, $AOB = \gamma$. Let OD be the diagonal required, and let OE be the diagonal of the face OAB. Then

$$\begin{aligned} OD^2 &= OE^2 + ED^2 + 2OE \,.\, ED\cos COE \\ &= a^2 + b^2 + 2ab\cos\gamma + c^2 + 2cOE\cos COE. \end{aligned}$$

Describe a sphere with O as centre meeting OA, OB, OC, OE at a, b, c, e respectively; then (see Example 14, Chap. IV.)

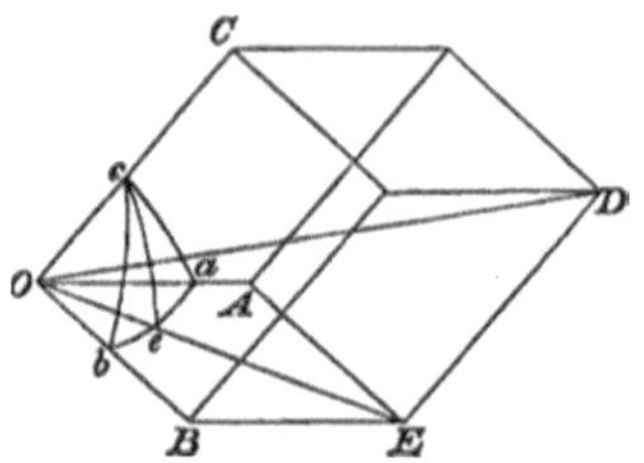

$$\cos cOe = \frac{\cos cOb \sin aOe + \cos cOa \sin bOe}{\sin aOb}$$

$$= \frac{\cos\alpha \sin aOe + \cos\beta \sin bOe}{\sin\gamma};$$

therefore

$$OD^2 = a^2 + b^2 + c^2 + 2ab\cos\gamma + \frac{2cOE}{\sin\gamma}(\cos\alpha \sin aOe + \cos\beta \sin bOe),$$

and $\qquad OE \sin aOe = b \sin\gamma, \qquad OE \sin bOe = a \sin\gamma;$

therefore $\quad OD^2 = a^2 + b^2 + c^2 + 2ab\cos\gamma + 2bc\cos\alpha + 2ca\cos\beta.$

158. *To find the volume of a tetrahedron.*

A tetrahedron is one-sixth of a parallelepiped which has the same altitude and its base double that of the tetrahedron; thus if the edges and their inclinations are given we can take one-sixth of the expression for the volume in Art. 156. The volume of a tetrahedron may also be expressed in terms of its six edges; for in the figure of Art. 156 let $BC = a'$, $CA = b'$, $AB = c'$; then

$$\cos\alpha = \frac{b^2 + c^2 - a'^2}{2bc}, \quad \cos\beta = \frac{c^2 + a^2 - b'^2}{2ca}, \quad \cos\gamma = \frac{a^2 + b^2 - c'^2}{2ab},$$

and if these values are substituted for $\cos\alpha$, $\cos\beta$, and $\cos\gamma$ in the expression obtained in Art. 156, the volume of the tetrahedron will be expressed in terms of its six edges.

The following result will be obtained, in which V denotes the volume of the tetrahedron,

$$144\, V^2 = -a'^2 b'^2 c'^2$$

$$+ a^2a'^2(b'^2 + c'^2 - a'^2) + b^2b'^2(c'^2 + a'^2 - b'^2) + c^2c'^2(a'^2 + b'^2 - c'^2)$$
$$- a'^2(a^2 - b^2)(a^2 - c^2) - b'^2(b^2 - c^2)(b^2 - a^2) - c'^2(c^2 - a^2)(c^2 - b^2).$$

Thus for a *regular* tetrahedron we have $144\, V^2 = 2a^6$.

159. If the vertex of a tetrahedron be supposed to be situated at any point in the plane of its base, the volume vanishes; hence if we equate to zero the expression on the right-hand side of the equation just given, we obtain a relation which must hold among the six straight lines which join four points taken arbitrarily in a plane.

Or we may adopt Carnot's method, in which this relation is established independently, and the expression for the volume of a tetrahedron is deduced from it; this we shall now shew, and we shall add some other investigations which are also given by Carnot.

It will be convenient to alter the notation hitherto used, by interchanging the accented and unaccented letters.

160. *To find the relation holding among the six straight lines which join four points taken arbitrarily in a plane.*

Let A, B, C, D be the four points. Let $AB = c$, $BC = a$, $CA = b$; also let $DA = a'$, $DB = b'$, $DC = c'$.

If D falls *within* the triangle ABC, the sum of the angles ADB, BDC, CDA is equal to four right angles; so that

$$\cos ADB = \cos(BDC + CDA).$$

Hence by ordinary transformations we deduce

$$1 = \cos^2 ADB + \cos^2 BDC + \cos^2 CDA - 2\cos ADB \cos BDC \cos CDA.$$

If D falls *without* the triangle ABC, one of the three angles at D is equal to the sum of the other two, and the result just given still holds.

Now $\cos ADB = \dfrac{a'^2 + b'^2 - c^2}{2a'b'}$, and the other cosines may be expressed in a similar manner; substitute these values in the above result, and we obtain the required relation, which after reduction may be exhibited thus,

$$\begin{aligned}0 &= -a^2b^2c^2\\ &+a'^2a^2(b^2+c^2-a^2)+b'^2b^2(c^2+a^2-b^2)+c'^2c^2(a^2+b^2-c^2)\\ &-a^2(a'^2-b'^2)(a'^2-c'^2)-b^2(b'^2-c'^2)(b'^2-a'^2)-c^2(c'^2-a'^2)(c'^2-b'^2).\end{aligned}$$

161. *To express the volume of a tetrahedron in terms of its six edges.*

Let a, b, c be the lengths of the sides of a triangle ABC forming one face of the tetrahedron, which we may call its base; let a', b', c' be the lengths of the straight lines which join A, B, C respectively to the vertex of the tetrahedron. Let p be the length of the perpendicular from the vertex on the base; then the lengths of the straight lines drawn from the foot of the perpendicular to A, B, C respectively are $\sqrt{(a'^2-p^2)}$, $\sqrt{(b'^2-p^2)}$, $\sqrt{(c'^2-p^2)}$. Hence the relation given in Art. 160 will hold if we put $\sqrt{(a'^2-p^2)}$ instead of a', $\sqrt{(b'^2-p^2)}$ instead of b', and $\sqrt{(c'^2-p^2)}$ instead of c'. We shall thus obtain

$$\begin{aligned}&p^2(2a^2b^2+2b^2c^2+2c^2a^2-a^4-b^4-c^4)=-a^2b^2c^2\\ &+a'^2a^2(b^2+c^2-a^2)+b'^2b^2(c^2+a^2-b^2)+c'^2c^2(a^2+b^2-c^2)\\ &-a^2(a'^2-b'^2)(a'^2-c'^2)-b^2(b'^2-c'^2)(b'^2-a'^2)-c^2(c'^2-a'^2)(c'^2-b'^2).\end{aligned}$$

The coefficient of p^2 in this equation is sixteen times the square of the area of the triangle ABC; so that the left-hand member is $144V^2$, where V denotes the volume of the tetrahedron. Hence the required expression is obtained.

162. *To find the relation holding among the six arcs of great circles which join four points taken arbitrarily on the surface of a sphere.*

Let A, B, C, D be the four points. Let $AB=\gamma$, $BC=\alpha$, $CA=\beta$; let $DA=\alpha'$, $DB=\beta'$, $DC=\gamma'$.

As in Art. 160 we have

$$1=\cos^2 ADB+\cos^2 BDC+\cos^2 CDA-2\cos ADB\cos BDC\cos CDA.$$

Now $\cos ADB=\dfrac{\cos\gamma-\cos\alpha'\cos\beta'}{\sin\alpha'\sin\beta'}$, and the other cosines may be expressed in a similar manner; substitute these values in the above result, and we obtain the required relation, which after reduction may be exhibited thus,

$$\begin{aligned}1 &= \cos^2\alpha + \cos^2\beta + \cos^2\gamma + \cos^2\alpha' + \cos^2\beta' + \cos^2\gamma' \\ &- \cos^2\alpha\cos^2\alpha' - \cos^2\beta\cos^2\beta' - \cos^2\gamma\cos^2\gamma' \\ &- 2(\cos\alpha\cos\beta\cos\gamma + \cos\alpha\cos\beta'\cos\gamma' \\ &+ \cos\beta\cos\alpha'\cos\gamma' + \cos\gamma\cos\alpha'\cos\beta') \\ &+ 2(\cos\alpha\cos\beta\cos\alpha'\cos\beta' + \cos\beta\cos\gamma\cos\beta'\cos\gamma' \\ &+ \cos\gamma\cos\alpha\cos\gamma'\cos\alpha').\end{aligned}$$

163. *To find the radius of the sphere circumscribing a tetrahedron.*

Denote the edges of the tetrahedron as in Art. 161. Let the sphere be supposed to be circumscribed about the tetrahedron, and draw on the sphere the six arcs of great circles joining the angular points of the tetrahedron. Then the relation given in Art. 162 holds among the cosines of these six arcs.

Let r denote the radius of the sphere. Then

$$\cos\alpha = 1 - 2\sin^2\frac{\alpha}{2} = 1 - 2\left(\frac{a}{2r}\right)^2 = 1 - \frac{a^2}{2r^2};$$

and the other cosines may be expressed in a similar manner. Substitute these values in the result of Art. 162, and we obtain, after reduction, with the aid of Art. 161,

$$4\times 144\,V^2r^2 = 2a^2b^2a'^2b'^2 + 2b^2c^2b'^2c'^2 + 2c^2a^2c'^2a'^2 - a^4a'^4 - b^4b'^4 - c^4c'^4.$$

The right-hand member may also be put into factors, as we see by recollecting the mode in which the expression for the area of a triangle is put into factors. Let $aa' + bb' + cc' = 2\sigma$; then

$$36\,V^2r^2 = \sigma(\sigma - aa')(\sigma - bb')(\sigma - cc').$$

EXAMPLES.

1. If I denote the inclination of two adjacent faces of a regular polyhedron, shew that $\cos I = \frac{1}{3}$ in the tetrahedron, $= 0$ in the cube, $= -\frac{1}{3}$ in the octahedron,

$= -\frac{1}{5}\sqrt{5}$ in the dodecahedron, and $= -\frac{1}{3}\sqrt{5}$ in the icosahedron.

2. With the notation of Art. 153, shew that the radius of the sphere which touches one face of a regular polyhedron and all the adjacent faces produced is $\frac{1}{2}a \cot \dfrac{\pi}{m} \cot \frac{1}{2}I$.

3. A sphere touches one face of a regular tetrahedron and the other three faces produced: find its radius.

4. If a and b are the radii of the spheres inscribed in and described about a regular tetrahedron, shew that $b = 3a$.

5. If a is the radius of a sphere inscribed in a regular tetrahedron, and R the radius of the sphere which touches the edges, shew that $R^2 = 3a^2$.

6. If a is the radius of a sphere inscribed in a regular tetrahedron, and R' the radius of the sphere which touches one face and the others produced, shew that $R' = 2a$.

7. If a cube and an octahedron be described about a given sphere, the sphere described about these polyhedrons will be the same; and conversely.

8. If a dodecahedron and an icosahedron be described about a given sphere, the sphere described about these polyhedrons will be the same; and conversely.

9. A regular tetrahedron and a regular octahedron are inscribed in the same sphere: compare the radii of the spheres which can be inscribed in the two solids.

10. The sum of the squares of the four diagonals of a parallelepiped is equal to four times the sum of the squares of the edges.

11. If with all the angular points of any parallelepiped as centres equal spheres be described, the sum of the intercepted portions of the parallelepiped will be equal in volume to one of the spheres.

12. A regular octahedron is inscribed in a cube so that the corners of the octahedron are at the centres of the faces of the cube: shew that the volume of the cube is six times that of the octahedron.

13. It is not possible to fill any given space with a number of regular polyhedrons of the same kind, except cubes; but this may be done by means of

tetrahedrons and octahedrons which have equal faces, by using twice as many of the former as of the latter.

14. A spherical triangle is formed on the surface of a sphere of radius ρ; its angular points are joined, forming thus a pyramid with the straight lines joining them with the centre: shew that the volume of the pyramid is

$$\tfrac{1}{3}\rho^3\sqrt{(\tan r \tan r_1 \tan r_2 \tan r_3)},$$

where r, r_1, r_2, r_3 are the radii of the inscribed and escribed circles of the triangle.

15. The angular points of a regular tetrahedron inscribed in a sphere of radius r being taken as poles, four equal small circles of the sphere are described, so that each circle touches the other three. Shew that the area of the surface bounded by each circle is $2\pi r^2\left(1-\dfrac{1}{\sqrt{3}}\right)$.

16. If O be any point within a spherical triangle ABC, the product of the sines of any two sides and the sine of the included angle

$$= \sin AO \sin BO \sin CO \Big\{ \cot AO \sin BOC + \cot BO \sin COA + \cot CO \sin AOB \Big\}.$$

XIV

ARCS DRAWN TO FIXED POINTS ON THE SURFACE OF A SPHERE.

164. In the present Chapter we shall demonstrate various propositions relating to the arcs drawn from any point on the surface of a sphere to certain fixed points on the surface.

165. ABC is a spherical triangle having all its sides quadrants, and therefore all its angles right angles; T is any point on the surface of the sphere: to shew that

$$\cos^2 TA + \cos^2 TB + \cos^2 TC = 1.$$

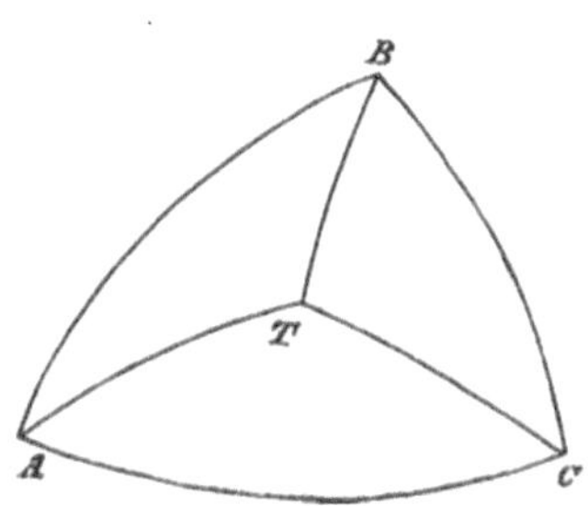

By Art. 37 we have

$$\cos TA = \cos AB \cos TB + \sin AB \sin TB \cos TBA$$
$$= \sin TB \cos TBA.$$

Similarly $\cos TC = \sin TB \cos TBC = \sin TB \sin TBA.$
Square and add; thus

$$\cos^2 TA + \cos^2 TC = \sin^2 TB = 1 - \cos^2 TB;$$

therefore $$\cos^2 TA + \cos^2 TB + \cos^2 TC = 1.$$

166. ABC is a spherical triangle having all its sides quadrants, and therefore all its angles right angles; T and U are any points on the surface of the sphere: to shew that

$$\cos TU = \cos TA \cos UA + \cos TB \cos UB + \cos TC \cos UC.$$

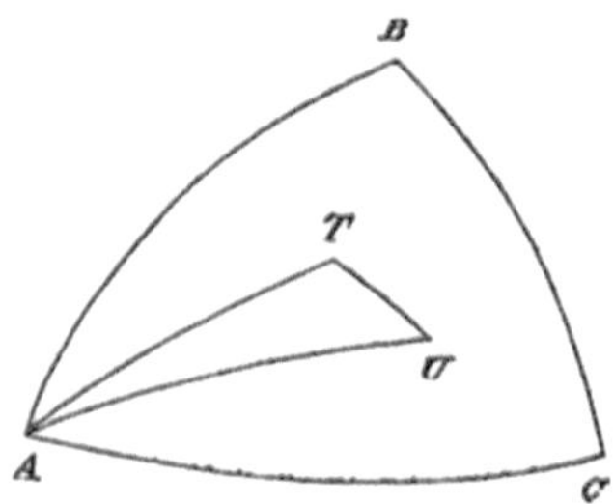

By Art. 37 we have

$$\cos TU = \cos TA \cos UA + \sin TA \sin UA \cos TAU,$$

and $$\cos TAU = \cos(BAU - BAT)$$
$$= \cos BAU \cos BAT + \sin BAU \sin BAT$$
$$= \cos BAU \cos BAT + \cos CAU \cos CAT;$$

therefore $$\cos TU = \cos TA \cos UA$$
$$+ \sin TA \sin UA(\cos BAU \cos BAT + \cos CAU \cos CAT);$$

and
$$\cos TB = \sin TA \cos BAT,$$
$$\cos UB = \sin UA \cos BAU,$$
$$\cos TC = \sin TA \cos CAT,$$
$$\cos UC = \sin UA \cos CAU;$$

therefore
$$\cos TU = \cos TA \cos UA + \cos TB \cos UB + \cos TC \cos UC.$$

167. We leave to the student the exercise of shewing that the formulæ of the two preceding Articles are perfectly general for all positions of T and U, outside or inside the triangle ABC: the demonstrations will remain essentially the same for all modifications of the diagrams. The formulæ are of constant application in Analytical Geometry of three dimensions, and are demonstrated in works on that subject; we have given them here as they may be of service in Spherical Trigonometry, and will in fact now be used in obtaining some important results.

168. Let there be any number of fixed points on the surface of a sphere; denote them by H_1, H_2, H_3,.... Let T be any point on the surface of a sphere. We shall now investigate an expression for the sum of the cosines of the arcs which join T with the fixed points.

Denote the sum by Σ; so that

$$\Sigma = \cos TH_1 + \cos TH_2 + \cos TH_3 + \ldots.$$

Take on the surface of the sphere a fixed spherical triangle ABC, having all its sides quadrants, and therefore all its angles right angles.

Let λ, μ, ν be the cosines of the arcs which join T with A, B, C respectively; let l_1, m_1, n_1 be the cosines of the arcs which join H_1 with A, B, C respectively; and let a similar notation be used with respect to H_2, H_3,....

Then, by Art. 166,

$$\Sigma = l_1\lambda + m_1\mu + n_1\nu + l_2\lambda + m_2\mu + n_2\nu + \ldots$$
$$= P\lambda + Q\mu + R\nu;$$

where P stands for $l_1 + l_2 + l_3 + \ldots$, with corresponding meanings for Q and R.

169. It will be seen that P is the value which Σ takes when T coincides with A, that Q is the value which Σ takes when T coincides with B, and that R is the value which Σ takes when T coincides with C. Hence the result expresses the general value of Σ in terms of the cosines of the arcs which join T to the fixed points A, B, C, and the particular values of Σ which correspond to these three points.

170. We shall now transform the result of Art. 168.

Let
$$G = \sqrt{(P^2 + Q^2 + R^2)};$$

and let α, β, γ be three arcs determined by the equations

$$\cos\alpha = \frac{P}{G}, \quad \cos\beta\frac{Q}{G}, \quad \cos\gamma\frac{R}{G};$$

then
$$\Sigma = G(\lambda\cos\alpha + \mu\cos\beta + \nu\cos\gamma).$$

Since $\cos^2\alpha + \cos^2\beta + \cos^2\gamma = 1$, it is obvious that there will be some point on the surface of the sphere, such that α, β, γ are the arcs which join it to A, B, C respectively; denote this point by U: then, by Art. 166,

$$\cos TU = \lambda\cos\alpha + \mu\cos\beta + \nu\cos\gamma;$$

and finally
$$\Sigma = G\cos TU.$$

Thus, whatever may be the position of T, the sum of the cosines of the arcs which join T to the fixed points varies as the cosine of the single arc which joins T to a certain fixed point U.

We might take G either positive or negative; it will be convenient to suppose it positive.

171. A sphere is described about a regular polyhedron; from any point on the surface of the sphere arcs are drawn to the solid angles of the polyhedron: to shew that the sum of the cosines of these arcs is zero.

From the preceding Article we see that if G is not zero there is *one* position of T which gives to Σ its greatest positive value, namely, when T coincides with U. But by the symmetry of a regular polyhedron there must always be *more than one* position of T which gives the same value to Σ. For instance, if we take a regular tetrahedron, as there are four faces there will at least be *three other* positions of T symmetrical with any assigned position.

Hence G must be zero; and thus the *sum of the cosines of the arcs which join* T *to the solid angles of the regular polyhedron is zero for all positions of* T.

172. Since $G = 0$, it follows that P, Q, R must each be zero; these indeed are particular cases of the general result of Art. 171. See Art. 169.

173. The result obtained in Art. 171 may be shewn to hold also in some other cases. Suppose, for instance, that a rectangular parallelepiped is inscribed in a sphere; then the sum of the cosines of the arcs drawn from any point on the surface of the sphere to the solid angles of the parallelepiped is zero. For here it is obvious that there must always be at least *one other* position of T symmetrical with any assigned position. Hence by the argument of Art. 171 we must have $G = 0$.

174. Let there be any number of fixed points on the surface of a sphere; denote them by H_1, H_2, $H_3, \ldots$ Let T be any point on the surface of the sphere. We shall now investigate a remarkable expression for the sum of the squares of the cosines of the arcs which join T with the fixed points.

Denote the sum by Σ; so that

$$\Sigma = \cos^2 TH_1 + \cos^2 TH_2 + \cos^2 TH_3 + \ldots.$$

Take on the surface of the sphere a fixed spherical triangle ABC, having all its sides quadrants, and therefore all its angles right angles.

Let λ, μ, ν be the cosines of the arcs which join T with A, B, C respectively; let l_1, m_1, n_1 be the cosines of the angles which join H_1 with A, B, C respectively; and let a similar notation be used with respect to H_2, $H_3, \ldots$.

Then, by Art. 166,

$$\Sigma = (l_1\lambda + m_1\mu + n_1\nu)^2 + (l_2\lambda + m_2\mu + n_2\nu)^2 + \ldots.$$

Expand each square, and rearrange the terms; thus

$$\Sigma = P\lambda^2 + Q\mu^2 + R\nu^2 + 2p\mu\nu + 2q\nu\lambda + 2r\lambda\mu,$$

where P stands for $l_1^2 + l_2^2 + l_3^2 + \dots,$

and p stands for $m_1 n_1 + m_2 n_2 + m_3 n_3 + \dots,$

with corresponding meanings for Q and q, and for R and r.

We shall now shew that there is some position of the triangle ABC for which p, q, and r will vanish; so that we shall then have

$$\Sigma = P\lambda^2 + Q\mu^2 + R\nu^2.$$

Since Σ is always a finite positive quantity there must be some position, or some positions, of T for which Σ has the largest value which it can receive. Suppose that A has this position, or one of these positions if there are more than one. When T is at A we have μ and ν each zero, and λ equal to unity, so that Σ is then equal to P.

Hence, whatever be the position of T, P is never less than $P\lambda^2 + Q\mu^2 + R\nu^2 + 2p\mu\nu + 2q\nu\lambda + 2r\lambda\mu$, that is, by Art. 165,

$P(\lambda^2 + \mu^2 + \nu^2)$ is never less than

$$P\lambda^2 + Q\mu^2 + R\nu^2 + 2p\mu\nu + 2q\nu\lambda + 2r\lambda\mu;$$

therefore

$(P - Q)\mu^2 + (P - R)\nu^2$ is never less than $2p\mu\nu + 2q\nu\lambda + 2r\lambda\mu$.

Now suppose $\nu = 0$; then T is situated on the great circle of which AB is a quadrant, and whatever be the position of T we have

$(P - Q)\mu^2$ not less than $2r\lambda\mu$,

and therefore $P - Q$ not less than $\dfrac{2r\lambda}{\mu}$.

But now $\dfrac{\lambda}{\mu}$ is equal to $\dfrac{\cos TA}{\cos TB}$; this is numerically equal to $\tan TB$, and so may be made numerically as great as we please, positive or negative, by giving a suitable position to T. Thus $P - Q$ must in some cases be less than $\frac{2r\lambda}{\mu}$ if r have any value different from zero.

Therefore r must $= 0$.

In like manner we can shew that q must $= 0$.

Hence with the specified position for A we arrive at the result that whatever may be the position of T

$$\Sigma = P\lambda^2 + Q\mu^2 + R\nu^2 + 2p\mu\,\nu.$$

Let us now suppose that the position of B is so taken that when T coincides with B the value of Σ is as large as it can be for any point in the great circle of which A is the pole. When T is at B we have λ and ν each zero, and μ equal to unity, so that Σ is then equal to Q. For any point in the great circle of which A is the pole λ is zero; and therefore for any such point

$$Q \text{ is not less than } Q\mu^2 + R\nu^2 + 2p\mu\,\nu,$$

that is, by Art. 165,

$$Q(\mu^2 + \nu^2) \text{ is not less than } Q\mu^2 + R\nu^2 + 2p\mu\,\nu;$$

therefore $Q - R$ is not less than $\frac{2p\mu}{\nu}$.

Hence by the same reasoning as before we must have $p = 0$.

Therefore we see that there must be some position of the triangle ABC, such that for every position of T

$$\Sigma = P\lambda^2 + Q\mu^2 + R\nu^2.$$

175. The remarks of Art. 169 are applicable to the result just obtained.

176. In the final result of Art. 174 we may shew that R is the least value which Σ can receive. For, by Art. 165,

$$\Sigma = P\lambda^2 + Q\mu^2 + R(1 - \lambda^2 - \mu^2)$$
$$= R + (P - R)\lambda^2 + (Q - R)\mu^2;$$

and by supposition neither $P - R$ nor $Q - R$ is negative, so that Σ cannot be less than R.

177. A sphere is described about a regular polyhedron; from any point on the surface of the sphere arcs are drawn to the solid angles of the polyhedron: it is required to find the sum of the squares of the cosines of these arcs.

With the notation of Art. 174 we have

$$\Sigma = P\lambda^2 + Q\mu^2 + R\nu^2.$$

We shall shew that in the present case P, Q, and R must *all be equal.* For if they are not, one of them must be greater than each of the others, or one of them must be less than each of the others.

If possible let the former be the case; suppose that P is greater than Q, and greater than R.

Now
$$\begin{aligned}\Sigma &= P(1 - \mu^2 - \nu^2) + Q\mu^2 + R\nu^2 \\ &= P - (P - Q)\mu^2 - (P - R)\nu^2;\end{aligned}$$

this shews that Σ is *always less than* P except when $\mu = 0$ and $\nu = 0$: that is Σ *is always less than* P except when T is at A, or at the point of the surface which is diametrically opposite to A. But by the symmetry of a regular polyhedron there must always be more than *two* positions of T which give the same value to Σ. For instance if we take a regular tetrahedron, as there are *four* faces there will be at least *three other* positions of T symmetrical with any assigned position. Hence P cannot be greater than Q and greater than R.

In the same way we can shew that one of the three P, Q, and R, cannot be less than each of the others.

Therefore $P = Q = R$; and therefore by Art. 165 for *every position* of T we have $\Sigma = P$.

Since $P = Q = R$ each of them $= \frac{1}{3}(P + Q + R)$

$$= \frac{1}{3}\{{l_1}^2 + {m_1}^2 + {n_1}^2 + {l_2}^2 + {m_2}^2 + {n_2}^2 + \ldots\}$$

$$= \frac{S}{3}, \text{ by Art. 165,}$$

where S is the number of the solid angles of the regular polyhedron.

Thus the sum of the squares of the cosines of the arcs which join any point on the surface of the sphere to the solid angles of the regular polyhedron is one third of the number of the solid angles.

178. Since $P = Q = R$ in the preceding Article, it will follow that when the fixed points of Art. 174 are the solid angles of a regular polyhedron, then for *any* position of the spherical triangle ABC we shall have $p = 0$, $q = 0$, $r = 0$.

For taking any position for the spherical triangle ABC we have

$$\Sigma = P\lambda^2 + Q\mu^2 + R\nu^2 + 2p\mu\nu + 2q\nu\lambda + 2r\lambda\mu;$$

then at A we have $\mu = 0$ and $\nu = 0$, so that P is then the value of Σ; similarly Q and R are the values of Σ at B and C respectively. But by Art. 177 we have the same value for Σ whatever be the position of T; thus

$$P = P(\lambda^2 + \mu^2 + \nu^2) + 2p\mu\nu + 2q\nu\lambda + 2r\lambda\mu;$$

therefore $$0 = 2p\mu\nu + 2q\nu\lambda + 2r\lambda\mu.$$

This holds then for every position of T. Suppose T is at *any point* of the great circle of which A is the pole; then $\lambda = 0$: thus we get $p\mu\nu = 0$, and therefore $p = 0$. Similarly $q = 0$, and $r = 0$.

179. Let there be any number of fixed points on the surface of a sphere; denote them by H_1, H_2, $H_3, \ldots$; from any two points T and U on the surface of the sphere arcs are drawn to the fixed points: it is required to find the sum of the products of the corresponding cosines, that is

$$\cos TH_1 \cos UH_1 + \cos TH_2 \cos UH_2 + \cos TH_3 \cos UH_3 + \ldots.$$

Let the notation be the same as in Art. 174; and let λ', μ', ν' be the cosines of the arcs which join U with A, B, C respectively. Then by Art. 166,

$$\cos TH_1 \cos UH_1 = (\lambda l_1 + \mu m_1 + \nu n_1)(\lambda' l_1 + \mu' m_1 + \nu' n_1) =$$
$$\lambda\lambda' {l_1}^2 + \mu\mu' {m_1}^2 + \nu\nu' {n_1}^2 + (\lambda\mu' + \mu\lambda')l_1 m_1 + (\mu\nu' + \nu\mu')m_1 n_1 + (\nu\lambda' + \lambda\nu')n_1 l_1.$$

Similar results hold for $\cos TH_2 \cos UH_2$, $\cos TH_3 \cos UH_3$,.... Hence, with the notation of Art. 174, the required sum is

$$\lambda\lambda' P + \mu\mu' Q + \nu\nu' R + (\mu\nu' + \nu\mu')p + (\nu\lambda' + \lambda\nu')q + (\lambda\mu' + \mu\lambda')r.$$

Now by properly choosing the position of the triangle ABC we have p, q, and r each zero as in Art. 174; and thus the required sum becomes

$$\lambda\lambda' P + \mu\mu' Q + \nu\nu' R.$$

180. The result obtained in Art. 174 may be considered as a particular case of that just given; namely the case in which the points T and U coincide.

181. A sphere is described about a regular polyhedron; from any two points on the surface of the sphere arcs are drawn to the solid angles of the polyhedron: it is required to find the sum of the products of the corresponding cosines.

With the notation of Art. 179 we see that the sum is

$$\lambda\lambda' P + \mu\mu' Q + \nu\nu' R.$$

And here $P = Q = R = \frac{S}{3}$, by Art. 177.

Thus the sum $= \frac{S}{3}(\lambda\lambda' + \mu\mu' + \nu\nu') = \frac{S}{3} \cos TU$.

Thus the sum of the products of the cosines is equal to the product of the cosine TU *into a third of the number of the solid angles of the regular polyhedron.*

182. The result obtained in Art. 177 may be considered as a particular case of that just given; namely, the case in which the points T and U coincide.

183. If TU is a quadrant then $\cos TU$ is zero, and the sum of the products of the cosines in Art. 181 is zero. The results $p = 0$, $q = 0$, $r = 0$, are easily seen to be all special examples of this particular case.

XV

MISCELLANEOUS PROPOSITIONS.

184. *To find the locus of the vertex of a spherical triangle of given base and area.*

Let AB be the given base, $= c$ suppose, $AC = \theta$, $BAC = \phi$. Since the area is given the spherical excess is known; denote it by E; then by Art. 103,

$$\cot \tfrac{1}{2}E = \cot \tfrac{1}{2}\theta \cot \tfrac{1}{2}c \operatorname{cosec} \phi + \cot \phi;$$

therefore
$$\sin(\phi - \tfrac{1}{2}E) = \cot \tfrac{1}{2}\theta \cot \tfrac{1}{2}c \sin \tfrac{1}{2}E;$$

therefore
$$2 \cot \tfrac{1}{2}c \sin \tfrac{1}{2}E \cos^2 \frac{\theta}{2} = \sin\theta \sin(\phi - \tfrac{1}{2}E);$$

therefore

$$\cos\theta \cot \tfrac{1}{2}c \sin \tfrac{1}{2}E + \sin\theta \cos\left(\phi - \tfrac{1}{2}E + \frac{\pi}{2}\right) = -\cot \tfrac{1}{2}c \sin \tfrac{1}{2}E.$$

Comparing this with equation (1) of Art. 133, we see that the required locus is a circle. If we call α, β the angular co-ordinates of its pole, we have

$$\tan\alpha = \frac{1}{\cot \frac{1}{2}c \sin \frac{1}{2}E} = \frac{\tan \frac{1}{2}c}{\sin \frac{1}{2}E},$$
$$\beta = \tfrac{1}{2}E - \frac{\pi}{2}.$$

It may be presumed from symmetry that the pole of this circle is in the great circle which bisects AB at right angles; and this presumption is easily verified. For the equation to that great circle is

$$0 = \cos\theta\cos\left(\frac{\pi}{2} - \frac{c}{2}\right) + \sin\theta\sin\left(\frac{\pi}{2} - \frac{c}{2}\right)\cos(\phi - \pi)$$

and the values $\theta = \alpha$, $\phi = \beta$ satisfy this equation.

185. *To find the angular distance between the poles of the inscribed and circumscribed circles of a triangle.*

Let P denote the pole of the inscribed circle, and Q the pole of the circumscribed circle of a triangle ABC; then $PAB = \frac{1}{2}A$, by Art. 89, and $QAB = S - C$, by Art. 92; hence

$$\cos PAQ = cos\tfrac{1}{2}(B - C);$$

$$\text{and } \cos PQ = \cos PA\cos QA + \sin PA\sin QA\cos\tfrac{1}{2}(B - C).$$

Now, by Art. 62 (see the figure of Art. 89),

$$\cos PA = \cos PE\cos AE = \cos r\cos(s - a),$$

$$\sin PA = \frac{\sin PE}{\sin PAE} = \frac{\sin r}{\sin\frac{1}{2}A};$$

thus

$$\cos PQ = \cos R\cos r\cos(s - a) + \sin R\sin r\cos\frac{1}{2}(B - C)\operatorname{cosec}\tfrac{1}{2}A.$$

Therefore, by Art. 54

$$\cos PQ = \cos R\cos r\cos(s - a) + \sin R\sin r\sin\tfrac{1}{2}(b + c)\operatorname{cosec}\tfrac{1}{2}a,$$

therefore $$\frac{\cos PQ}{\cos R\sin r} = \cot r\cos(s - a) + \tan R\sin\frac{1}{2}(b + c)\operatorname{cosec}\frac{1}{2}a.$$

Now $$\cot r = \frac{\sin s}{n}, \quad \tan R = \frac{2\sin\frac{1}{2}a\sin\frac{1}{2}b\sin\frac{1}{2}c}{n},$$

therefore $$\frac{\cos PQ}{\cos R \sin r} = \frac{1}{n}\left\{\sin s \cos(s-a) + 2\sin\tfrac{1}{2}(b+c)\sin\tfrac{1}{2}b\sin\tfrac{1}{2}c\right\}$$
$$= \frac{1}{2n}(\sin a + \sin b + \sin c).$$

Hence $$\left(\frac{\cos PQ}{\cos R \sin r}\right)^2 - 1 = \frac{1}{4n^2}(\sin a + \sin b + \sin c)^2 - 1$$
$$= (\cot r + \tan R)^2 \text{ (by Art. 94);}$$

therefore $$\cos^2 PQ = \cos^2 R \sin^2 r + \cos^2(R - r),$$

and $$\sin^2 PQ = \sin^2(R - r) - \cos^2 R \sin^2 r.$$

186. *To find the angular distance between the pole of the circumscribed circle and the pole of one of the escribed circles of a triangle.*

Let Q denote the pole of the circumscribed circle, and Q_1 the pole of the escribed circle opposite to the angle A. Then it may be shewn that $QBQ_1 = \frac{1}{2}\pi + \frac{1}{2}(C - A)$, and

$$\cos QQ_1 = \cos R \cos r_1 \cos(s-c) - \sin R \sin r_1 \sin\tfrac{1}{2}(C-A)\sec\tfrac{1}{2}B$$
$$= \cos R \cos r_1 \cos(s-c) - \sin R \sin r_1 \sin\tfrac{1}{2}(c-a)\,\mathrm{cosec}\,\frac{1}{2}b.$$

Therefore

$$\frac{\cos QQ_1}{\sin r_1 \cos R} = \cot r_1 \cos(s-c) - \tan R \sin\tfrac{1}{2}(c-a)\,\mathrm{cosec}\,\tfrac{1}{2}b;$$

by reducing as in the preceding Article, the right-hand member of the last equation becomes
$$\frac{1}{2n}(\sin b + \sin c - \sin a);$$

hence $$\left(\frac{\cos QQ_1}{\cos R \sin r_1}\right)^2 - 1 = (\tan R - \cot r_1)^2, \text{ (Art. 94);}$$

therefore $$\cos^2 QQ_1 = \cos^2 R \sin^2 r_1 + \cos^2(R + r_1),$$

and $$\sin^2 QQ_1 = \sin^2(R + r_1) - \cos^2 R \sin^2 r_1.$$

187. *The arc which passes through the middle points of the sides of any triangle upon a given base will meet the base produced at a fixed point, the distance of* *which from the middle point of the base is a quadrant.*

Let ABC be any triangle, E the middle point of AC, and F the middle point of AB; let the arc which joins E and F when produced meet BC produced at Q. Then

$$\frac{\sin BQ}{\sin BF} = \frac{\sin BFQ}{\sin BQF}, \qquad \frac{\sin AQ}{\sin AF} = \frac{\sin AFQ}{\sin AQF};$$

therefore
$$\frac{\sin BQ}{\sin AQ} = \frac{\sin AQF}{\sin BQF},$$

similarly
$$\frac{\sin CQ}{\sin AQ} = \frac{\sin AQF}{\sin CQF};$$

therefore $\qquad \sin BQ = \sin CQ; \qquad$ therefore $BQ + CQ = \pi$.

Hence if D be the middle point of BC

$$DQ = \tfrac{1}{2}(BQ + CQ) = \tfrac{1}{2}\pi.$$

188. *If three arcs be drawn from the angles of a spherical triangle through any point to meet the opposite sides, the products of the sines of the alternate segments of the sides are equal.*

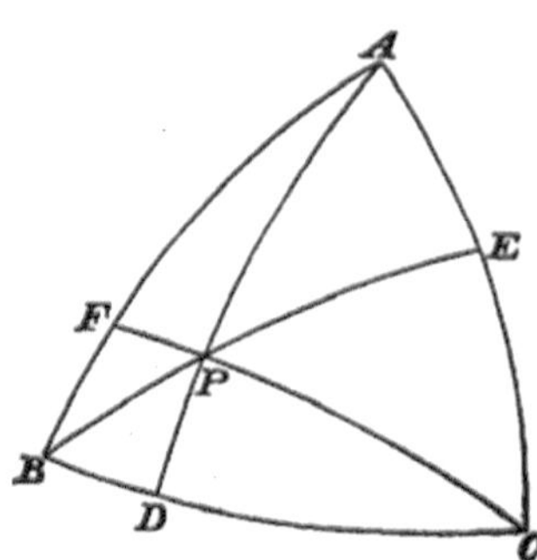

Let P be any point, and let arcs be drawn from the angles A, B, C passing through P and meeting the opposite sides at D, E, F. Then

$$\frac{\sin BD}{\sin BP} = \frac{\sin BPD}{\sin BDP}, \qquad \frac{\sin CD}{\sin CP} = \frac{\sin CPD}{\sin CDP},$$

therefore
$$\frac{\sin BD}{\sin CD} = \frac{\sin BPD}{\sin CPD}\,\frac{\sin BP}{\sin CP}.$$

Similar expressions may be found for $\frac{\sin CE}{\sin AE}$ and $\frac{\sin AF}{\sin BF}$; and hence it follows obviously that

$$\frac{\sin BD}{\sin CD}\,\frac{\sin CE}{\sin AE}\,\frac{\sin AF}{\sin BF} = 1;$$

therefore $$\sin BD \sin CE \sin AF = \sin CD \sin AE \sin BF.$$

189. Conversely, when the points D, E, F in the sides of a spherical triangle are such that the relation given in the preceding Article holds, the arcs which join these points with the opposite angles respectively *pass through a common point.* Hence the following propositions may be established: the perpendiculars from the angles of a spherical triangle on the opposite sides meet at a point; the arcs which bisect the angles of a spherical triangle meet at a point; the arcs which join the angles of a spherical triangle with the middle points of the opposite sides meet at a point; the arcs which join the angles of a spherical triangle with the points where the inscribed circle touches the opposite sides respectively meet at a point.

Another mode of establishing such propositions has been exemplified in Arts. 139 and 140.

190. *If* AB *and* A′B′ *be any two equal arcs* AA′ *and* AA′ *and* BB′ *be bisected at right angles by arcs meeting at* P, *then* AB *and* A′B′ *subtend equal angles at* P.

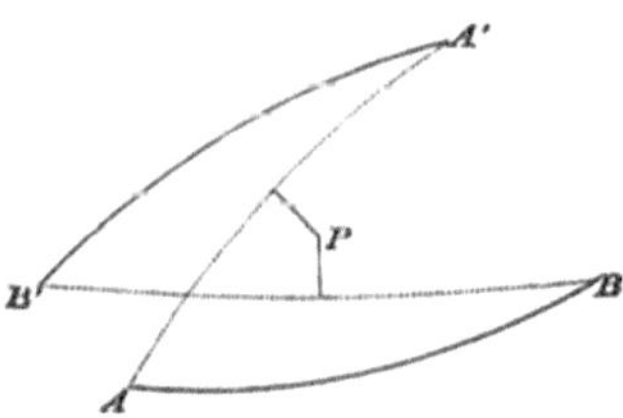

For $PA = PA'$ and $PB = PB'$; hence the sides of the triangle PAB are respectively equal to those of $PA'B'$; therefore the angle APB = the angle $A'PB'$.

This simple proposition has an important application to the motion of a rigid body of which one point is fixed. For conceive a sphere capable of motion round its centre which is fixed; then it appears from this proposition that any two fixed points on the sphere, as A and B, can be brought into any other positions, as A' and B', by rotation round an axis passing through the centre of the sphere and a certain point P. Hence it may be inferred that any change of position in a rigid body, of which one point is fixed, may be effected by rotation round some axis through the fixed point.

(De Morgan's *Differential and Integral Calculus*, page 489.)

191. Let P denote any point within any *plane* angle AOB, and from P draw perpendiculars on the straight lines OA and OB; then it is evident that these perpendiculars include an angle which is the *supplement* of the angle AOB. The corresponding fact with respect to a *solid* angle is worthy of notice. Let there be a solid angle formed by three plane angles, meeting at a point O. From any point P within the solid angle, draw perpendiculars PL, PM, PN on the three planes which form the solid angle; then the spherical triangle which corresponds to the three planes LPM, MPN, NPL is the *polar triangle* of the spherical triangle which corresponds to the solid angle at O. This remark is due to Professor De Morgan.

192. Suppose three straight lines to meet at a point and form a solid angle; let α, β, and γ denote the angles contained by these three straight lines taken in pairs: then it has been proposed to call the expression $\sqrt{(1 - \cos^2 \alpha - \cos^2 \beta - \cos^2 \gamma + 2 \cos \alpha \cos \beta \cos \gamma)}$, the *sine of the solid angle*. See Baltzer's *Theorie... der Determinanten*, 2nd edition, page 177. Adopting this definition it is easy to shew that the sine of a solid angle lies between zero and unity.

We know that the area of a plane triangle is half the product of two sides into the sine of the included angle: by Art. 156 we have the following analogous proposition; the volume of a tetrahedron is one sixth of the product of three edges into the sine of the solid angle which they form.

Again, we know in mechanics that if three forces acting at a point are in equilibrium, each force is as the sine of the angle between the directions of the other two: the following proposition is analogous; if four forces acting at a point are in equilibrium each force is as the sine of the solid angle formed by the directions of the other three. See *Statics*, Chapter II.

193. Let a sphere be described about a regular polyhedron; let perpendiculars be drawn from the centre of the sphere on the faces of the polyhedron, and produced to meet the surface of the sphere: then it is obvious from symmetry that the points of intersection must be the angular points of another regular polyhedron.

This may be verified. It will be found on examination that if S be the number of solid angles, and F the number of faces of one regular polyhedron, then another regular polyhedron exists which has S faces and F solid angles. See Art. 151.

194. *Polyhedrons.* The result in Art. 150 was first obtained by Euler; the demonstration which is there given is due to Legendre. The demonstration shews that the result is true in many cases in which the polyhedron has *re-entrant* solid angles; for all that is necessary for the demonstration is, that it shall be possible to take a point within the polyhedron as the centre of a sphere, so that the polygons, formed as in Art. 150, shall not have any coincident portions. The result, however, is generally true, even in cases in which the condition required by the demonstration of Art. 150 is not satisfied. We shall accordingly give another demonstration, and shall then deduce some important consequences from the result. We begin with a theorem which is due to Cauchy.

195. *Let there be any network of rectilineal figures, not necessarily in one plane, but not forming a closed surface; let* E *be the number of edges,* F *the number of figures, and* S *the number of corner points: then* F + S = E + 1.

This theorem is obviously true in the case of a single plane figure; for then $F = 1$, and $S = E$. It can be shewn to be generally true by induction. For assume the theorem to be true for a network of F figures; and suppose that a rectilineal figure of n sides is added to this network, so that the network and the additional figure have m sides coincident, and therefore $m + 1$ corner points coincident. And with respect to the new network which is thus formed, let E', F', S' denote the same things as E, F, S with respect to the old network. Then

$$E' = E + n - m, \quad F' = F + 1, \quad S' = S + n - (m + 1);$$

therefore $$F' + S' - E' = F + S - E.$$

But $F + S = E + 1$, by hypothesis; therefore $F' + S' = E' + 1$.

196. To demonstrate Euler's theorem we suppose one face of a polyhedron removed, and we thus obtain a network of rectilineal figures to which Cauchy's theorem is applicable. Thus

$$F - 1 + S = E + 1;$$

therefore

$$F + S = E + 2.$$

197. *In any polyhedron the number of faces with an odd number of sides is even, and the number of solid angles formed with an odd number of plane angles is even.*

Let a, b, c, d,...... denote respectively the numbers of faces which are triangles, quadrilaterals, pentagons, hexagons,....... Let α, β, γ, δ,...... denote respectively the numbers of the solid angles which are formed with three, four, five, six,...... plane angles.

Then, each edge belongs to *two* faces, and terminates at *two* solid angles; therefore

$$2E = 3a + 4b + 5c + 6d + \ldots\ldots,$$

$$2E = 3\alpha + 4\beta + 5\gamma + 6\delta + \ldots\ldots.$$

From these relations it follows that $a + c + e + \ldots\ldots$, and $\alpha + \gamma + \epsilon + \ldots\ldots$ are *even* numbers.

198. With the notation of the preceding Article we have

$$F = a + b + c + d + \ldots\ldots,$$

$$S = \alpha + \beta + \gamma + \delta + \ldots\ldots.$$

From these combined with the former relations we obtain

$$2E - 3F = b + 2c + 3d + \ldots\ldots,$$

$$2E - 3S = \beta + 2\gamma + 3\delta + \ldots\ldots.$$

Thus $2E$ cannot be less than $3F$, or less than $3S$.

199. From the expressions for E, F, and S, given in the two preceding Articles, combined with the result $2F + 2S = 4 + 2E$, we obtain

$$2(a+b+c+d+\ldots)+2(\alpha+\beta+\gamma+\delta+\ldots)=4+3a+4b+5c+6d+\ldots,$$

$$2(a+b+c+d+\ldots)+2(\alpha+\beta+\gamma+\delta+\ldots)=4+3\alpha+4\beta+5\gamma+6\delta+\ldots,$$

therefore
$$2(\alpha+\beta+\gamma+\delta+\ldots)-(a+2b+3c+4d+\ldots)=4, \qquad (1)$$
$$2(a+b+c+d+\ldots)-(\alpha+2\beta+3\gamma+4\delta+\ldots)=4. \qquad (2)$$

Therefore, by addition

$$a+\alpha-(c+\gamma)-2(d+\delta)-3(e+\epsilon)-\ldots\ldots=8.$$

Thus the number of triangular faces together with the number of solid angles formed with three plane angles cannot be less than eight.

Again, from (1) and (2), by eliminating α, we obtain

$$3a+2b+c-e-2f-\ldots\ldots-2\beta-4\gamma-\ldots\ldots=12,$$

so that $3a+2b+c$ cannot be less than 12. From this result various inferences can be drawn; thus for example, *a solid cannot be formed which shall have no triangular, quadrilateral, or pentagonal faces.*

In like manner, we can shew that $3\alpha+2\beta+\gamma$ cannot be less than 12.

200. Poinsot has shewn that in addition to the five well-known *regular polyhedrons*, four other solids exist which are perfectly symmetrical in shape, and which might therefore also be called *regular*. We may give an idea of the nature of Poinsot's results by referring to the case of a polygon. Suppose five points A, B, C, D, E, placed in succession at equal distances round the circumference of a circle. If we draw a straight line from each point to the next point, we form an ordinary regular pentagon. Suppose however we join the points by straight lines in the following order, A to C, C to E, E to B, B to D, D to A; we thus form a star-shaped symmetrical figure, which might be considered a regular pentagon.

It appears that, in a like manner, four, and only four, new regular solids can be formed. To such solids, the faces of which intersect and cross, Euler's theorem does not apply.

201. Let us return to Art. 195, and suppose e the number of edges *in* the bounding contour, and e' the number of edges *within* it; also suppose s the number of corners *in* the bounding contour, and s' the number *within* it. Then

$$E = e + e'; \; S = s + s';$$

therefore $$1 + e + e' = s + s' + F.$$

But $$e = s;$$

therefore $$1 + e' = s' + F.$$

We can now demonstrate an extension of Euler's theorem, which has been given by Cauchy.

202. *Let a polyhedron be decomposed into any number of polyhedrons at pleasure; let* P *be the number thus formed,* S *the number of solid angles,* F *the number of faces,* E *the number of edges: then* S + F = E + P + 1.

For suppose all the polyhedrons united, by starting with one and adding one at a time. Let e, f, s be respectively the numbers of edges, faces, and solid angles in the first; let e', f', s' be respectively the numbers of edges, faces, and solid angles in the second which are not common to it and the first; let e'', f'', s'' be respectively the numbers of edges, faces, and solid angles in the third which are not common to it and the first or second; and so on. Then we have the following results, namely, the first by Art. 196, and the others by Art. 201;

$$s + f = e + 2,$$
$$s' + f' = e' + 1,$$
$$s'' + f'' = e'' + 1,$$
$$\dots\dots\dots\dots\dots\dots$$

By addition, since $s + s' + s'' + \dots = S$, $f + f' + f'' + \dots = F$, and $e + e' + e'' + \dots = E$, we obtain

$$S + F = E + P + 1.$$

203. The following references will be useful to those who study the theory of polyhedrons. Euler, *Novi Commentarii Academiæ... Petropolitanæ*, Vol. IV.

1758; Legendre, *Géométrie*; Poinsot, *Journal de l'École Polytechnique*, Cahier X; Cauchy, *Journal de l'École Polytechnique*, Cahier XVI; Poinsot and Bertrand, *Comptes Rendus... de l'Académie des Sciences*, Vol. XLVI; Catalan, *Théorèmes et Problèmes de Géométrie Elémentaire*; Kirkman, *Philosophical Transactions* for 1856 and subsequent years; Listing, *Abhandlungen der Königlichen Gesellschaft... zu Göttingen*, Vol. X.

MISCELLANEOUS EXAMPLES.

1. Find the locus of the vertices of all right-angled spherical triangles having the same hypotenuse; and from the equation obtained, prove that the locus is a circle when the radius of the sphere is infinite.

2. AB is an arc of a great circle on the surface of a sphere, C its middle point: shew that the locus of the point P, such that the angle APC = the angle BPC, consists of two great circles at right angles to one another. Explain this when the triangle becomes plane.

3. On a given arc of a sphere, spherical triangles of equal area are described: shew that the locus of the angular point opposite to the given arc is defined by the equation

$$\tan^{-1}\left\{\frac{\tan(\alpha+\phi)}{\sin\theta}\right\}+\tan^{-1}\left\{\frac{\tan(\alpha-\phi)}{\sin\theta}\right\}+\tan^{-1}\left\{\frac{\tan\theta}{\sin(\alpha+\phi)}\right\}+\tan^{-1}\left\{\frac{\tan\theta}{\sin(\alpha-\phi)}\right\}=\beta,$$

where 2α is the length of the given arc, θ the arc of the great circle drawn from any point P in the locus perpendicular to the given arc, ϕ the inclination of the great circle on which θ is measured to the great circle bisecting the given arc at right angles, and β a constant.

4. In any spherical triangle

$$\tan c=\frac{\cot A\cot a+\cot B\cot b}{\cot a\cot b-\cos A\cos B}.$$

5. If θ, ϕ, ψ denote the distances from the angles A, B, C respectively of the point of intersection of arcs bisecting the angles of the spherical triangle ABC,

shew that

$$\cos\theta\sin(b-c)+\cos\phi\sin(c-a)+\cos\psi\sin(a-b)=0.$$

6. If A', B', C' be the poles of the sides BC, CA, AB of a spherical triangle ABC, shew that the great circles AA', BB', CC' meet at a point P, such that

$$\cos PA\cos BC=\cos PB\cos CA=\cos PC\cos AB.$$

7. If O be the point of intersection of arcs AD, BE, CF drawn from the angles of a triangle perpendicular to the opposite sides and meeting them at D, E, F respectively, shew that

$$\frac{\tan AD}{\tan OD},\quad \frac{\tan BE}{\tan OE},\quad \frac{\tan CF}{\tan OF}$$

are respectively equal to

$$1+\frac{\cos A}{\cos B\cos C},\quad 1+\frac{\cos B}{\cos A\cos C},\quad 1+\frac{\cos C}{\cos A\cos B}.$$

8. If p, q, r be the arcs of great circles drawn from the angles of a triangle perpendicular to the opposite sides, (α,α'), (β,β'), (γ,γ') the segments into which these arcs are divided, shew that

$$\tan\alpha\tan\alpha'=\tan\beta\tan\beta'=\tan\gamma\tan\gamma';$$

and
$$\frac{\cos p}{\cos\alpha\cos\alpha'}=\frac{\cos q}{\cos\beta\cos\beta'}=\frac{\cos r}{\cos\gamma\cos\gamma'}.$$

9. In a spherical triangle if arcs be drawn from the angles to the middle points of the opposite sides, and if α, α' be the two parts of the one which bisects the side a, shew that

$$\frac{\sin\alpha}{\sin\alpha'}=2\cos\frac{a}{2}.$$

10. The arc of a great circle bisecting the sides AB, AC of a spherical triangle cuts BC produced at Q: shew that

$$\cos AQ\sin\frac{a}{2}=\sin\frac{c-b}{2}\sin\frac{c+b}{2}.$$

11. If $ABCD$ be a spherical quadrilateral, and the opposite sides AB, CD when produced meet at E, and AD, BC meet at F, the ratio of the sines of the arcs drawn from E at right angles to the diagonals of the quadrilateral is the same as the ratio of those from F.

12. If $ABCD$ be a spherical quadrilateral whose sides AB, DC are produced to meet at P, and AD, BC at Q, and whose diagonals AC, BD intersect at R, then

$$\sin AB \sin CD \cos P = \sin AD \sin BC \cos Q = \sin AC \sin BD \cos R.$$

13. If A' be the angle of the chordal triangle which corresponds to the angle A of a spherical triangle, shew that

$$\cos A' = \sin(S - A) \cos \frac{a}{2}.$$

14. If the tangent of the radius of the circle described about a spherical triangle is equal to twice the tangent of the radius of the circle inscribed in the triangle, the triangle is equilateral.

15. The arc AP of a circle of the same radius as the sphere is equal to the greater of two sides of a spherical triangle, and the arc AQ taken in the same direction is equal to the less; the sine PM of AP is divided at E, so that $\frac{EM}{PM}$ = the natural cosine of the angle included by the two sides, and EZ is drawn parallel to the tangent to the circle at Q. Shew that the remaining side of the spherical triangle is equal to the arc QPZ.

16. If through any point P within a spherical triangle ABC great circles be drawn from the angular points A, B, C to meet the opposite sides at a, b, c respectively, prove that

$$\frac{\sin Pa \cos PA}{\sin Aa} + \frac{\sin Pb \cos PB}{\sin Bb} + \frac{\sin Pc \cos PC}{\sin Cc} = 1.$$

17. A and B are two places on the Earth's surface on the same side of the equator, A being further from the equator than B. If the bearing of A from B be more nearly due East than it is from any other place in the same latitude as B, find the bearing of B from A.

18. From the result given in example 18 of Chapter V. infer the possibility

of a regular dodecahedron.

19. A and B are fixed points on the surface of a sphere, and P is any point on the surface. If a and b are given constants, shew that a fixed point S can always be found, in AB or AB produced, such that

$$a \cos AP + b \cos BP = s \cos SP,$$

where s is a constant.

20. A, B, C,... are fixed points on the surface of a sphere; a, b, c,... are given constants. If P be a point on the surface of the sphere, such that

$$a \cos AP + b \cos BP + c \cos CP + \ldots = \text{constant},$$

shew that the locus of P is a circle.

XVI

NUMERICAL SOLUTION OF SPHERICAL TRIANGLES.

204. We shall give in this Chapter examples of the numerical solution of Spherical Triangles.

We shall first take right-angled triangles, and then oblique-angled triangles.

Right-Angled Triangles.

205. Given $a = 37^\circ\, 48'\, 12''$, $b = 59^\circ\, 44'\, 16''$, $C = 90^\circ$.

To find c we have

$$
\begin{aligned}
\cos c &= \cos a \cos b, \\
L \cos 37^\circ\, 48'\, 12'' &= \quad 9.8976927 \\
L \cos 59^\circ\, 44'\, 16'' &= \quad \underline{9.7023945} \\
L \cos c + 10 &= \quad 19.6000872 \\
c &= \quad 66^\circ\, 32'\, 6''.
\end{aligned}
$$

To find A we have

$$\cot A = \cot a \sin b,$$

$$\begin{aligned} L\cot 37^\circ\, 48'\, 12'' &= 10.1102655 \\ L\sin 59^\circ\, 44'\, 16'' &= 9.9363770 \\ \hline L\cot A + 10 &= 20.0466425 \\ A &= 41^\circ\, 55'\, 45''. \end{aligned}$$

To find B we have

$$\cot B = \cot b \sin a,$$

$$\begin{aligned} L\cot 59^\circ\, 44'\, 16'' &= 9.7660175 \\ L\sin 37^\circ\, 48'\, 12'' &= 9.7874272 \\ \hline L\cot B + 10 &= 19.5534447 \\ B &= 70^\circ\, 19'\, 15''. \end{aligned}$$

206. Given $A = 55^\circ\, 32'\, 45''$, $C = 90^\circ$, $c = 98^\circ\, 14'\, 24''$.

To find a we have

$$\sin a = \sin c \sin A,$$

$$\begin{aligned} L\sin 98^\circ\, 14'\, 24'' &= 9.9954932 \\ L\sin 55^\circ\, 32'\, 45'' &= 9.9162323 \\ \hline L\sin a + 10 &= 19.9117255 \\ a &= 54^\circ\, 41'\, 35''. \end{aligned}$$

To find B we have

$$\cot B = \cos c \tan A.$$

Here $\cos c$ *is negative*; and therefore $\cot B$ will be negative, and B greater than a right angle. The numerical value of $\cos c$ is the same as that of $\cos 81^\circ\, 45'\, 36''$.

$$\begin{aligned} L\cos 81^\circ\, 45'\, 36'' &= 9.1563065 \\ L\tan 55^\circ\, 32'\, 45'' &= 10.1636102 \\ \hline L\cot(180^\circ - B) + 10 &= 19.3199167 \\ 180^\circ - B &= 78^\circ\, 12'\, 4'' \\ B &= 101^\circ\, 47'\, 56''. \end{aligned}$$

To find b we have

$$\tan b = \tan c \cos A.$$

Here $\tan c$ is *negative*; and therefore $\tan b$ will be negative and b greater than a quadrant.

$$\begin{aligned} L\tan 81^\circ\, 45'\, 36'' &= 10.8391867 \\ L\cos 55^\circ\, 32'\, 45'' &= 9.7526221 \\ \hline L\tan(180^\circ - b) + 10 &= 20.5918088 \\ 180^\circ - b &= 75^\circ 38'\, 32'' \\ b &= 104^\circ\, 21'\, 28''. \end{aligned}$$

207. Given $A = 46^\circ\, 15'\, 25''$, $C = 90^\circ$, $a = 42^\circ\, 18'\, 45''$.

To find c we have

$$\sin c = \frac{\sin a}{\sin A},$$

$$L\sin c = 10 + L\sin a - L\sin A,$$

$$\begin{aligned} 10 + L\sin 42^\circ\, 18'\, 45'' &= 19.8281272 \\ L\sin 46^\circ\, 15'\, 25'' &= 9.8588065 \\ \hline L\sin c &= 9.9693207 \\ c = 68^\circ\, 42'\, 59'' \text{ or } & 111^\circ\, 17'\, 1''. \end{aligned}$$

To find b we have

$$\sin b = \tan a \cot A,$$

$$\begin{aligned} L\tan 42^\circ\, 18'\, 45'' &= 9.9591983 \\ L\cot 46^\circ\, 15'\, 25'' &= 9.9809389 \\ \hline L\sin b + 10 &= 19.9401372 \\ b = 60^\circ\, 36'\, 10'' \text{ or } & 119^\circ\, 23'\, 50''. \end{aligned}$$

To find B we have

$$\sin B = \frac{\cos A}{\cos a},$$

$$L\sin B = 10 + L\cos A - L\cos a,$$

$$\begin{aligned} 10 + L\cos 46^\circ\, 15'\, 25'' &= 19.8397454 \\ L\cos 42^\circ\, 18'\, 45'' &= 9.8689289 \\ \hline L\sin B &= 9.9708165 \\ B = 69^\circ\, 13'\, 47'' \text{ or } & 110^\circ\, 46'\, 13''. \end{aligned}$$

Oblique-Angled Triangles.

208. Given $a = 70^\circ\, 14'\, 20''$, $b = 49^\circ\, 24'\, 10''$, $c = 38^\circ\, 46'\, 10''$. We shall use the formula given in Art. 45,

$$\tan \tfrac{1}{2}A = \sqrt{\left\{\frac{\sin(s-b)\sin(s-c)}{\sin s \sin(s-a)}\right\}}.$$

Here
$$s = 79^\circ\, 12'\, 20'',$$
$$s - a = 8^\circ\, 58',$$
$$s - b = 29^\circ\, 48'\, 10'',$$
$$s - c = 40^\circ\, 26'\, 10''.$$

$$\begin{array}{rr}
L\sin 29^\circ\, 48'\, 10'' = & 9.6963704 \\
L\sin 40^\circ\, 26'\, 10'' = & 9.8119768 \\
\hline
 & 19.5083472 \\
L\sin 79^\circ\, 12'\, 20'' = & 9.9922465 \\
L\sin 8^\circ\, 58' = & 9.1927342 \\
\hline
 & 19.1849807 \\
 & 19.5083472 \\
 & 19.1849807 \\
\hline
 & 2\,)\,.3233665 \\
L\tan\tfrac{1}{2}A - 10 = & .1616832
\end{array}$$

$$\tfrac{1}{2}A = 55^\circ\, 25'\, 38''$$
$$A = 110^\circ\, 51'\, 16''.$$

Similarly to find B,

$$
\begin{aligned}
L\sin 8^\circ\,58' &= 9.1927342\\
L\sin 40^\circ\,26'\,10'' &= 9.8119768\\
\hline
&\ 19.0047110\\
L\sin 79^\circ\,12'\,20'' &= 9.9922465\\
L\sin 29^\circ\,48'\,10'' &= 9.6963704\\
\hline
&\ 19.6886169\\
&\ 19.0047110\\
&\ 19.6886169\\
\hline
&\ 2\,)\ \bar{1}.3160941\\
L\tan\tfrac{1}{2}B - 10 &= \bar{1}.6580470\\
L\tan\tfrac{1}{2}B &= 9.6580470\\
\tfrac{1}{2}B &= 24^\circ\,28'\,2''\\
B &= 48^\circ\,56'\,4''.
\end{aligned}
$$

Similarly to find C,

$$\begin{aligned}
L\sin 8^\circ\, 58' &= 9.1927342 \\
L\sin 29^\circ\, 48'\, 10'' &= 9.6963704 \\
&\overline{18.8891046} \\
L\sin 79^\circ\, 12'\, 20'' &= 9.9922465 \\
L\sin 40^\circ\, 26'\, 10'' &= 9.8119768 \\
&\overline{19.8042233} \\
&18.8891046 \\
&19.8042233 \\
&\overline{2\,)\ \bar{1}.0848813} \\
L\tan\tfrac{1}{2}C - 10 &= \bar{1}.5424406 \\
L\tan\tfrac{1}{2}C &= 9.5424406 \\
\tfrac{1}{2}C &= 19^\circ\, 13'\, 24'' \\
C &= 38^\circ\, 26'\, 48''.
\end{aligned}$$

209. Given $a = 68^\circ\, 20'\, 25''$, $b = 52^\circ\, 18'\, 15''$, $C = 117^\circ\, 12'\, 20''$.

By Art. 82,

$$\tan\tfrac{1}{2}(A+B) = \frac{\cos\frac{1}{2}(a-b)}{\cos\frac{1}{2}(a+b)}\cot\tfrac{1}{2}C,$$

$$\tan\frac{1}{2}(A-B) = \frac{\sin\frac{1}{2}(a-b)}{\sin\frac{1}{2}(a+b)}\cot\tfrac{1}{2}C.$$

$$\tfrac{1}{2}(a-b) = 8^\circ\, 1'\, 5'',\ \tfrac{1}{2}(a+b) = 60^\circ\, 19'\, 20'',\ \tfrac{1}{2}C = 58^\circ\, 36'\, 10''.$$

$$\begin{aligned}
L\cos 8^\circ\, 1'\, 5'' &= 9.9957335 \\
L\cot 58^\circ\, 36'\, 10'' &= 9.7855690 \\
\hline
&\ 19.7813025 \\
L\cos 60^\circ\, 19'\, 20'' &= 9.6947120 \\
\hline
L\tan \tfrac{1}{2}(A+B) &= 10.0865905 \\
\tfrac{1}{2}(A+B) &= 50^\circ\, 40'\, 28'' \\
L\sin 8^\circ\, 1'\, 5'' &= 9.1445280 \\
L\cot 58^\circ\, 36'\, 10'' &= 9.7855690 \\
\hline
&\ 18.9300970 \\
L\sin 60^\circ\, 19'\, 20'' &= 9.9389316 \\
\hline
L\tan \tfrac{1}{2}(A-B) &= 8.9911654 \\
\tfrac{1}{2}(A-B) &= 5^\circ\, 35'\, 47''.
\end{aligned}$$

Therefore $\qquad A = 56^\circ\, 16'\, 15'', \quad B = 45^\circ\, 4'\, 41''.$

If we proceed to find c from the formula

$$\sin c = \frac{\sin a \sin C}{\sin A},$$

since $\sin C$ is greater than $\sin A$ we shall obtain two values for c both greater than a, and we shall not know which is the value to be taken.

We shall therefore determine c from formula (1) of Art. 54, which is free from ambiguity,

$$\cos \tfrac{1}{2}c - \frac{\cos \frac{1}{2}(a+b)\sin \frac{1}{2}C}{\cos \frac{1}{2}(A+B)},$$

$$\begin{aligned}
L\cos 60^\circ\, 19'\, 20'' &= 9.6947120 \\
L\sin 58^\circ\, 36'\, 10'' &= 9.9312422 \\
\hline
&\ 19.6259542 \\
L\cos 50^\circ\, 40'\, 28'' &= 9.8019015 \\
\hline
L\cos \tfrac{1}{2}c &= 9.8240527 \\
\tfrac{1}{2}c &= 48^\circ\, 10'\, 22'' \\
c &= 96^\circ\, 20'\, 44''.
\end{aligned}$$

Or we may adopt the second method of Art. 82. First, we determine θ from the formula $\tan\theta = \tan b \cos C$.

Here $\cos C$ *is negative*, and therefore $\tan\theta$ will be negative, and θ greater than a right angle. The numerical value of $\cos C$ is the same as that of $\cos 62^\circ\, 47'\, 40''$.

$$
\begin{aligned}
L\tan 52^\circ\, 18'\, 15'' &= 10.1119488 \\
L\cos 62^\circ\, 47'\, 40'' &= 9.6600912 \\
\hline
L\tan(180^\circ - \theta) + 10 &= 19.7720400 \\
180^\circ - \theta &= 30^\circ\, 36'\, 33'', \\
\text{therefore } \theta &= 149^\circ\, 23'\, 27''.
\end{aligned}
$$

Next, we determine c from the formula

$$\cos c = \frac{\cos b \cos(a - \theta)}{\cos\theta}.$$

Here $\cos\theta$ *is negative*, and therefore $\cos c$ will be negative, and c will be greater than a right angle. The numerical value of $\cos\theta$ is the same as that of $\cos(180^\circ - \theta)$, that is, of $\cos 30^\circ\, 36'\, 33''$; and the value of $\cos(a - \theta)$ is the same as that of $\cos(\theta - a)$, that is, of $\cos 81^\circ\, 3'\, 2''$.

$$
\begin{aligned}
L\cos 52^\circ\, 18'\, 15'' &= 9.7863748 \\
L\cos 81^\circ\, 3'\, 2'' &= 9.1919060 \\
\hline
& \ 18.9782808 \\
L\cos 30^\circ\, 36'\, 33'' &= 9.9348319 \\
\hline
L\cos(180^\circ - c) &= 9.0434489 \\
180^\circ - c &= 83^\circ\, 39'\, 17'' \\
c &= 96^\circ\, 20'\, 43''.
\end{aligned}
$$

Thus by taking only the nearest number of seconds in the tables the two methods give values of c which differ by $1''$; if, however, we estimate fractions of a second both methods will agree in giving about $43\frac{1}{2}$ as the number of seconds.

210. Given $a = 50^\circ\, 45'\, 20''$, $b = 69^\circ\, 12'\, 40''$, $A = 44^\circ\, 22'\, 10''$.

By Art. 84,
$$\sin B = \frac{\sin b}{\sin a}\sin A,$$

$$
\begin{aligned}
L\sin 69^\circ\, 12'\, 40'' &= 9.9707626 \\
L\sin 44^\circ\, 22'\, 10'' &= 9.8446525 \\
\hline
&\ 19.8154151 \\
L\sin 50^\circ\, 45'\, 20'' &= 9.8889956 \\
\hline
L\sin B &= 9.9264195 \\
B &= 57^\circ\, 34'\, 51''.4, \text{ or } 122^\circ\, 25'\, 8''.6.
\end{aligned}
$$

In this case there will be two solutions; see Art. 86. We will calculate C and c by Napier's analogies,

$$\tan \tfrac{1}{2}C = \frac{\cos \frac{1}{2}(b-a)}{\cos \frac{1}{2}(b+a)} \cot \tfrac{1}{2}(B+A),$$

$$\tan \tfrac{1}{2}c = \frac{\cos \frac{1}{2}(B+A)}{\cos \frac{1}{2}(B-A)} \tan \tfrac{1}{2}(b+a).$$

First take the smaller value of B; thus

$$\tfrac{1}{2}(B+A) = 50^\circ\, 58'\, 30''.7,\ \tfrac{1}{2}(B-A) = 6^\circ\, 35'\, 20''.7,$$

$$
\begin{aligned}
L\cos 9^\circ\, 13'\, 40'' &= 9.9943430\\
L\cot 50^\circ\, 58'\, 30''.7 &= 9.9087536\\
&\overline{19.9030966}\\
L\cos 59^\circ\, 59' &= 9.6991887\\
L\tan \tfrac{1}{2}C &= 10.2039079\\
\tfrac{1}{2}C &= 57^\circ\, 58'\, 55''.3\\
C &= 115^\circ\, 57'\, 50''.6.
\end{aligned}
$$

$$
\begin{aligned}
L\cos 50^\circ\, 58'\, 30''.7 &= 9.7991039\\
L\tan 59^\circ\, 59' &= 10.2382689\\
&\overline{20.0373728}\\
L\cos 6^\circ\, 36'\, 20''.7 &= 9.9971072\\
L\tan \tfrac{1}{2}c &= 10.0402656\\
\tfrac{1}{2}c &= 47^\circ\, 39'\, 8''.2\\
c &= 95^\circ\, 18'\, 16''.4.
\end{aligned}
$$

Next take the larger value of B; thus

$$\tfrac{1}{2}(B+A) = 83^\circ\, 23'\, 39''.3, \quad \tfrac{1}{2}(B-A) = 39^\circ\, 1'\, 29''.3.$$

$$
\begin{aligned}
L\cos\ 9^\circ\,13'\,40'' &= \ \ 9.9943430 \\
L\cot 83^\circ\,23'\,39''.3 &= \ \ 9.0637297 \\
&\quad\ \overline{19.0580727} \\
L\cos 59^\circ\,59' &= \ \ 9.6991887 \\
L\tan\tfrac{1}{2}C &= \ \ 9.3588840 \\
\tfrac{1}{2}C &= 12^\circ\,52'\,15''.8 \\
C &= 25^\circ\,44'\,31''.6.
\end{aligned}
$$

$$
\begin{aligned}
L\cos 83^\circ\,23'\,39''.3 &= \ \ 9.0608369 \\
L\tan 59^\circ\,59' &= 10.2382689 \\
&\quad\ \overline{19.2991058} \\
L\cos 39^\circ\,1'\,29''.3 &= \ \ 9.8903494 \\
L\tan\tfrac{1}{2}c &= \ \ 9.4087564 \\
\tfrac{1}{2}c &= 14^\circ\,22'\,32''.6 \\
c &= 28^\circ\,45'\,5''.2
\end{aligned}
$$

The student can obtain more examples, which can be easily verified, from those here worked out, by interchanging the given and required quantities, or by making use of the polar triangle.

EXAMPLES.

1. Given $b = 137^\circ\,3'\,48''$, $A = 147^\circ\,2'\,54''$, $C = 90^\circ$.

   Results. $c = 47^\circ\,57'\,15''$, $a = 156^\circ\,10'\,34''$, $B = 113^\circ\,28'$.

2. Given $c = 61^\circ\,4'\,56''$, $a = 40^\circ\,31'\,20''$, $C = 90^\circ$.

   Results. $b = 50^\circ\,30'\,29''$, $B = 61^\circ\,50'\,28''$, $A = 47^\circ\,54'\,21''$.

3. Given $A = 36^\circ$, $B = 60^\circ$, $C = 90^\circ$.

   Results. $a = 20^\circ\,54'\,18''.5$, $b = 31^\circ\,43'\,3''$, $c = 37^\circ\,21'\,38''.5$.

4. Given $a = 59^\circ\,28'\,27''$, $A = 66^\circ\,7'\,20''$, $C = 90^\circ$.

   Results. $c = 70^\circ\,23'\,42''$, $b = 48^\circ\,39'\,16''$, $B = 52^\circ\,50'\,20''$, or, $c = 109^\circ\, 36'\, 18''$, $b = 131^\circ\, 20'\, 44''$, $B = 127^\circ\, 9'\, 40''$.

5. Given $c = 90^\circ$, $a = 138^\circ\, 4'$, $b = 109^\circ\, 41'$.

Results. $C = 113^\circ\, 28'\, 2''$, $A = 142^\circ\, 11'\, 38''$, $B = 120^\circ\, 15'\, 57''$.

6. Given $c = 90^\circ$, $A = 131^\circ\, 30'$, $B = 120^\circ\, 32'$.

Results. $C = 109^\circ\, 40'\, 20''$, $a = 127^\circ\, 17'\, 51''$, $b = 113^\circ\, 49'\, 31''$.

7. Given $a = 76^\circ\, 35'\, 36''$, $b = 50^\circ\, 10'\, 30''$, $c = 40^\circ\, 0'\, 10''$.

Results. $A = 121^\circ\, 36'\, 20''$, $B = 42^\circ\, 15'\, 13''$, $C = 34^\circ\, 15'\, 3''$.

8. Given $A = 129^\circ\, 5'\, 28''$, $B = 142^\circ\, 12'\, 42''$, $C = 105^\circ\, 8'\, 10''$.

Results. $a = 135^\circ\, 49'\, 20''$, $b = 144^\circ\, 37'\, 15''$, $c = 60^\circ\, 4'\, 54''$.

www.ingramcontent.com/pod-product-compliance
Lightning Source LLC
LaVergne TN
LVHW101914190826
846094LV00001B/3

Let FPP' be the straight line, and draw Ax parallel to it. Join FS, and find the points D and E such that

$$SD : DF :: SE : EF :: SA : Ax.$$

Describe the circle on DE as diameter, and let it intersect the given line in P and P'.

Join DP, EP and draw SG, FH at right angles to EP.

Then DPE, being the angle in a semicircle, is a right angle, and DP is parallel to SG and FH.

Hence
$$\begin{aligned} SG : FH &:: SE : EF \\ &:: SD : DF \\ &:: PG : PH; \end{aligned}$$

therefore the angles SPG, FPH are equal, and therefore PD bisects the angle SPF.

Hence $SP : PF :: SD : DF :: SA : Ax$,
and P is a point in the curve.

Similarly P' is also a point in the curve, and the perpendicular from O, the centre of the circle, on FPP' meets it in V, the middle point of the chord PP'.

Since
$$SE : EF :: SA : Ax$$
and
$$SD : DF :: SA : Ax;$$
$$\therefore SE - SD : DE :: SA : Ax,$$
or
$$SO : OD :: SA : Ax,$$
a relation analogous to
$$SC : AC :: SA : AX.$$

We have already shewn, for each conic, that the middle points of parallel chords lie in a straight line; the following article contains a proof of the theorem which includes all the three cases.

185. Prop. II. *To find the locus of the middle points of a system of parallel chords.*

Let $P'P$ one of the chords be produced to meet the directrix in F, draw Ax parallel to FP, and divide FS so that

$$SD : DF :: SE : EF :: SA : Ax;$$

then, as in the preceding article, the perpendicular OV upon PP' from O, the middle point of DE, bisects PP'.

Draw the parallel focal chord aSa'; then Oc parallel to the directrix bisects aa' in c. Also draw SG perpendicular to the chords, and meeting the directrix in G.

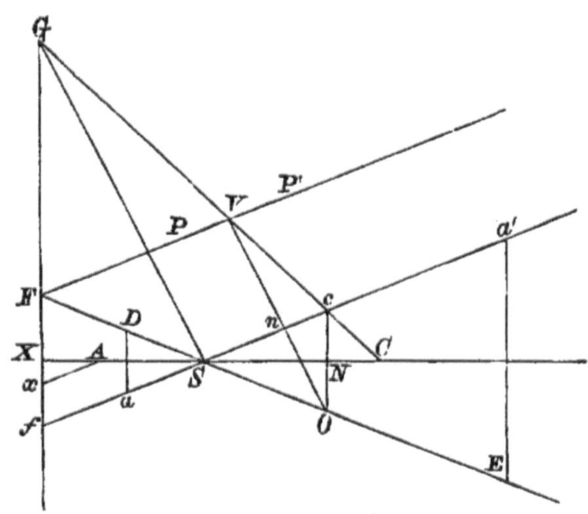

Then, if OV meet aa' in n,

$$Vn : nO :: SF : SO,$$
$$:: Sf : Sc,$$

and, since ncO, SGf are similar triangles,

$$nO : nc :: SG : Sf;$$
$$\therefore Vn : nc :: SG : Sc,$$

and the line Vc passes through G.

The straight line Gc is therefore the locus of the middle points of all chords parallel to aSa'.

The ends of the diameter GC may be found by the construction of the preceding article.

186. When the conic is a parabola, $SA = AX$,

and $$Sa : af :: AX : Ax$$
$$:: SX : Sf.$$

So $$Sa' : a'f :: SX : Sf;$$

$$\therefore Sc : ac :: SX : Sf,$$

and $$ac : cf :: SX : Sf.$$

Hence $$Sc : cf :: SX^2 : Sf^2$$
$$:: GX . Xf : Gf . fX$$
$$:: GX : Gf;$$

and therefore Gc is parallel to SX, that is, the middle points of parallel chords of a parabola lie in a straight line parallel to the axis.

187. PROP. III. *To find the locus of the middle points of all focal chords of a conic.*

Taking the case of a central conic, and referring to the figure of the preceding article, let Oc meet SC in N;

then $$cN : NS :: fX : SX,$$

and $$cN : NC :: GX : CX;$$

$$\therefore cN^2 : SN . NC :: fX . GX : SX . CX$$
$$:: SX^2 : SX . CX.$$

Hence it follows that the locus of c is an ellipse of which SC is the transverse axis, and such that the squares of its axes are as $SX : CX$, or (Cor. Art. 63) as $BC^2 : AC^2$.

Hence the locus of c is similar to the conic itself.

EXAMPLES.

1. If an ordinate, PNP', to the transverse axis meet the tangent at the end of the latus rectum in T,

$$SP = TN, \text{ and } TP . TP' = SN^2.$$

2. A focal chord PSQ of a conic section is produced to meet the directrix in K, and KM, KN are drawn through the feet of the ordinates PM, QN of P and Q. If KN produced meet PN produced in R, prove that

$$PR = PM.$$

3. The tangents at P and Q, two points in a conic, intersect in T; if through P, Q, chords be drawn parallel to the tangents at Q and P, and intersecting the conic in p and q respectively, and if tangents at p and q meet in T, shew that Tt is a diameter.

4. Two tangents TP, TP' are drawn to a conic intersecting the directrix in F, F'.

If the chord PQ cut the directrix in R, prove that

$$SF : SF' :: RF : RF'.$$

5. The chord of a conic PP' meets the directrix in K, and the tangents at P and P' meet in T; if RKR', parallel to ST, meet the tangents in R and R',

$$KR = KR'.$$

6. The tangents at P and P', intersecting in T, meet the latus rectum in D and D'; prove that the lines through D and D', respectively perpendicular to SP and SP', intersect in ST.

7. If P, Q be two points on a conic, and p, q two points on the directrix such that pq subtends at the focus half the angle subtended by PQ, either Pp and Qq or Pq and Qp meet on the curve.

8. A chord PP' of a conic meets the directrix in F, and from any point T in PP', TLL' is drawn parallel to SF and meeting SP, SP' in L and L'; prove that the ratio of SL or SL' to the distance of T from the directrix is equal to the ratio of $SA : AX$.

9. If an ellipse and an hyperbola have their axes coincident and proportional, points on them equidistant from one axis have the sum of the squares on their distances from the other axis constant.

10. If Q be any point in the normal PG, QR the perpendicular on SP, and QM the perpendicular on PN,

$$QR : PM :: SA : AX.$$

11. Given a focus of a conic section inscribed in a triangle, find the points where it touches the sides.

12. PSQ is any focal chord of a conic section; the normals at P and Q intersect in K, and KN is drawn perpendicular to PQ; prove that PN is equal to SQ, and hence deduce the locus of N.

13. Through the extremity P, of the diameter PQ of an ellipse, the tangent TPT' is drawn meeting two conjugate diameters in T, T'. From P, Q the lines PR, QR are drawn parallel to the same conjugate diameters. Prove that the rectangle under the semi-axes of the ellipse is a mean proportional between the triangles PQR and CTT'.

14. Shew that a conic may be drawn touching the sides of a triangle, having one focus at the centre of the circumscribing circle, and the other at the orthocentre.

15. The perpendicular from the focus of a conic on any tangent, and the central radius to the point of contact, intersect on the directrix.

16. AB, AC are tangents to a conic at B, and C, and $DEGF$ is drawn from a point D in AC, parallel to AB and cutting the curve in E and F, and BC in G; shew that

$$DG^2 = DE \,.\, DF.$$

17. A diameter of a parabola, vertex F, meets two tangents in D and E and their chord of contact is G, shew that

$$FG^2 = ED \,.\, FE.$$

18. P and Q are two fixed points in a parabola, and from any other point R in the curve, RP, RQ are drawn cutting a fixed diameter, vertex E, in B and C; prove that the ratio of EB to EC is constant.

19. If the normal at P meet the conjugate axis in g, and gk be perpendicular to SP, Pk is constant; and if kl, parallel to the transverse axis, meet the normal at P in l, kl is constant.

20. A system of conics is drawn having a common focus S and a common latus rectum LSL'. A fixed straight line through S intersects the conics, and at the points of intersection normals are drawn. Prove that the envelope of each of these normals is a parabola whose focus lies on LSL', and which has the given line as tangent at the vertex.

CHAPTER X.

ELLIPSES AS ROULETTES AND GLISSETTES.

188. *If a circle rolls on the inside of the circumference of a circle of double its radius, any point in the area of the rolling circle traces out an ellipse.*

Let C be the centre of the rolling circle, E the point of contact.

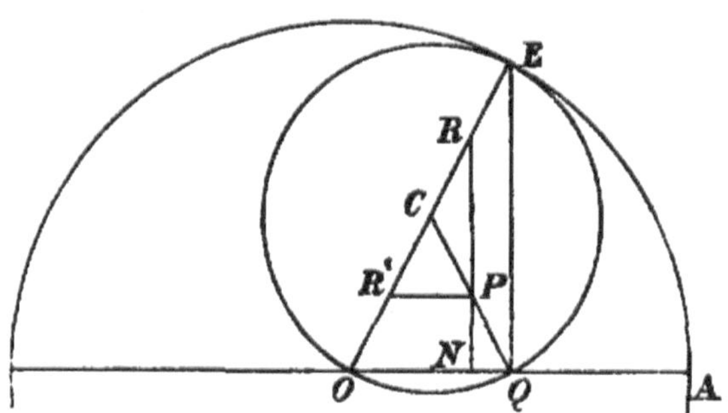

Then, if the circle meet in Q a fixed radius OA of the fixed circle, the angle ECQ is twice the angle EOA, and therefore the arcs EQ, EA are equal.

Hence, when the circles touch at A, the point Q of the rolling circle coincides with A, and the subsequent path of Q is the diameter through A.

Let P be a given point in the given radius CQ, and draw RPN perpendicular to OA, and PR' parallel to OA.

Then, OQE being a right angle, EQ is parallel to RP and therefore $CR = CP = CR'$, so that OR and OR' are constant.

Also $$PN : RN :: PQ : OR;$$

therefore, the locus of R being a circle, the locus of P is an ellipse, whose axes are as $PQ : OR$.

But OR is clearly the length of one semi-axis, and PQ or OR' is therefore the length of the other, OR, OR' being equal to $OC + CP$ and $OC - CP$.

189. Properties of the ellipse are deducible from this construction.

Thus, as the circle rolls, the point E is instantaneously at rest, and the motion of P is therefore at right angles to EP, *i.e.* producing EP to F, in the direction FO.

Therefore, drawing PT parallel to OF, PT is the tangent, and PF the normal.

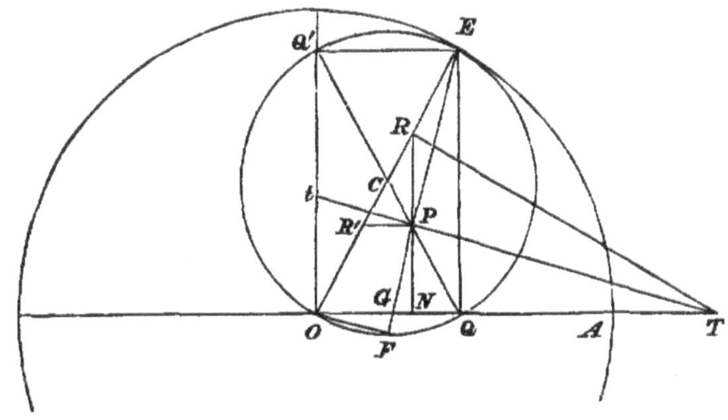

The angles EPT, EQT being right angles, the points E, P, Q, T are concyclic; but the circle through QPE clearly passes through R; therefore the angle ERT and consequently the angle ORT is a right angle,

and $$ON : OR :: OR : OT,$$

or $$ON \,.\, OT = OR^2,$$

which is the theorem of Art. 74.

Again, since $EQ't$ and EPt are right angles, E, Q', t, P are concyclic; but the circle through $EQ'P$ clearly passes through R'; therefore the angle $ER't$ and consequently the angle $OR't$ is a right angle, and

$$PN : OR' :: OR' : Ot,$$

or $$PN \,.\, Ot = OR'^2,$$

which is the theorem of Art. 75.

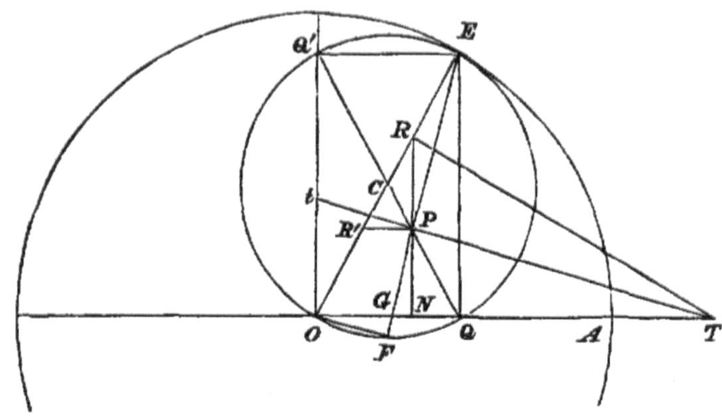

Further, if PF meet OQ in G, the angles PQG, PFQ are equal, being on equal bases EQ, OQ';

$$\therefore PG : PQ :: PQ : PF,$$

or
$$PG \,.\, PF = PQ^2 = OR'^2,$$

which is the first of theorems of Art. 77.

And again, if PGF produced meet $Q'O$ produced in g, the angles $PQ'g$, PFQ' are equal, being on equal bases QO, EQ'; and the angle $Q'Pg$ is common to the two triangles $PQ'g$, PFQ'.

Therefore these triangles are similar, and

$$Pg : PQ' :: PQ' : PF,$$

or
$$Pg \,.\, PF = PQ'^2.$$

But
$$PQ' = ER' = OR;$$

$$\therefore Pg \,.\, PF = OR^2,$$

which is the second theorem of Art. 77.

190. If the carried point P is outside the circle the line PNR, perpendicular to OA, will meet OE produced in R, and CR will be equal to CP, so that OR will be constant and the locus of R will be a circle.

Also, the triangles PQN, RON being similar, we shall have

$$PN : RN :: PQ : OR,$$

so that the locus of P will be an ellipse, the semi-axes of which will be $CP + OC$ and $CP - OC$.

191. The fact that a point on the circumference of the rolling circle oscillates in a straight line is utilized in the construction of Wheatstone's Photometer.

By help of machinery a metallic circle, about an inch in diameter, is made to roll rapidly round the inside of a circle of double this diameter, and carries a small bright bead which is fastened to its circumference.

If this machine is held between two candles or other sources of light, so that the line of oscillation of the bead is equidistant from the candles, two bright lines will be seen in close contiguity, and it is easy to form an estimate of their comparative brightnesses.

If bright beads are fastened to points in the area of the rolling circle not on the circumference, and the machine be held near sources of light, the appearance, when the circle is made to rotate rapidly, will be that of a number of bright concentric ellipses.

192. *A given straight line has its ends moveable on two straight lines at right angles to each other; the path of any given point in the moving line is an ellipse.*

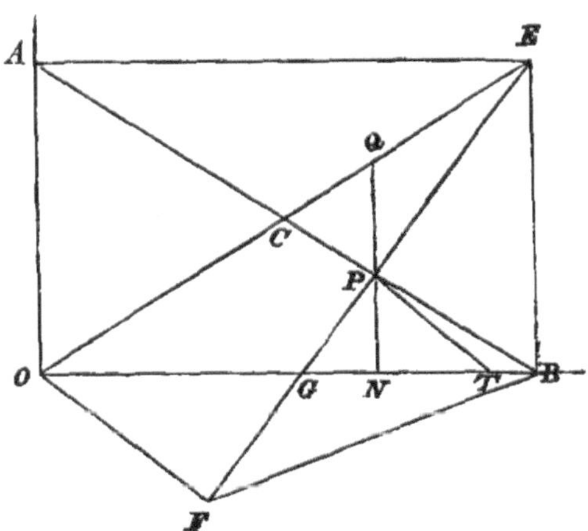

Let P be the point in the moving line AB, and C the middle point of AB.

Let the ordinate NP, produced if necessary, meet OC in Q; then $CQ = CP$ and $OQ = AP$, so that the locus of Q is a circle.

Also
$$PN : QN :: PB : OQ$$
$$:: PB : PA;$$

therefore the locus of P is an ellipse, and its semi-axes are equal to AP and BP.

193. The theorem of Art. 188 is at once reducible to this case, for, taking the figure of Art. 189, QPQ' is a diameter of the rolling circle and is therefore of constant length, and the points Q and Q' move along fixed straight lines at right angles to each other; the locus of P is therefore an ellipse of which $Q'P$ and PQ are the semi-axes.

194. From this construction also properties of the tangent and normal are deducible.

Complete the rectangle $OAEB$; then, since the directions of motion of A and B are respectively perpendicular to EA and EB, the state of motion of the line AB may be represented by supposing that the triangle EAB is turning round the point E.

Hence it follows that EP is the normal to the locus of P, and that PT perpendicular to EP is the tangent.

Let OF, parallel to PT, meet EP in F; then O, F, B, E are concyclic;

$$\therefore \text{ the angle } PFB = EOB = PBG,$$

and the triangles PGB, PFB are similar.

Hence $$PG : PB :: PB : PF,$$

or $$PG \,.\, PF = PB^2,$$

where PB is equal to the semi-conjugate axis.

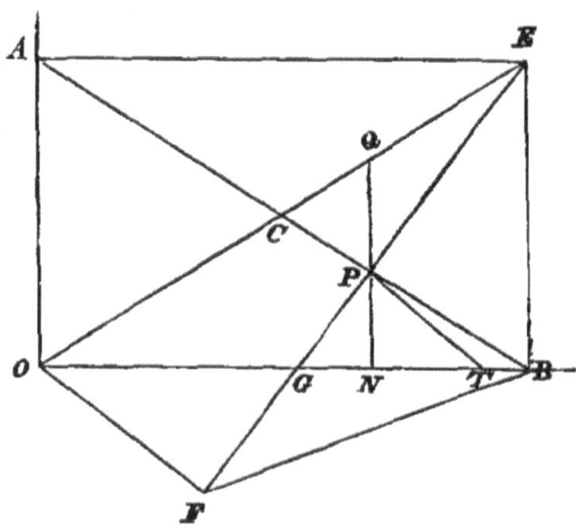

Similarly, by joining AF, it can be shewn that

$$Pg \,.\, PF = PA^2,$$

g being the point of intersection of PG and AO.

Again, since EPT, EBT are right angles, B, T, P, E are concyclic, and Q is clearly concyclic with B, P, E; so that TQE is a right angle.

Hence OQN and OQT are similar triangles, and

$$ON : OQ :: OQ : OT,$$

or

$$ON \,.\, OT = PA^2,$$

where PA is equal to the semi-transverse axis.

195. Observing that F, O, A, E, B are concyclic, we have

$$PF \,.\, PE = PA \,.\, PB;$$

$\therefore PE$ is equal to the semi-diameter conjugate to OP.

This suggests a construction for the solution of the problem,

Having given a pair of conjugate diameters of an ellipse, it is required to determine the position and magnitudes of the principal axes.

Taking OP and OD as the given semi-conjugate diameters, draw PF perpendicular to OD, and, in FP produced, take PE equal to OD.

Join OE, bisect it in C, and in CE take CQ equal to CP.

Then OB, OA, drawn perpendicular and parallel to PQ, and meeting CP in B and A, will be the directions of the axes, and their lengths will be AP and PB.

196. *If a given triangle AQB move in its own plane so that the extremities A, B, of its base AB move on two fixed straight lines at right angles to each other, the path of the point Q is an ellipse.*

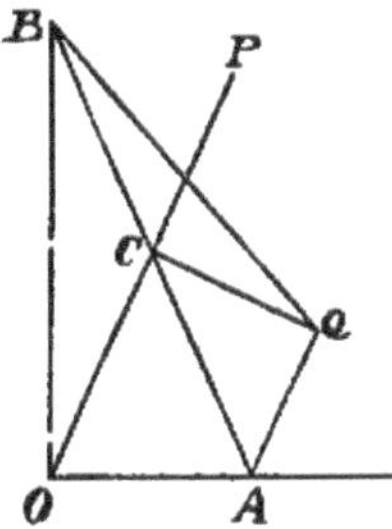

If O be the point of intersection of the fixed lines, and C the middle point of AB, the angles COB, CBO are equal, so that, as AB slides, the line CB, and therefore also the line CQ, turns round as fast as CO, but in the contrary direction.

Produce OC to P, making $CP = CQ$; then the locus of P is a circle the radius of which is equal to $OC + CQ$.

There is clearly one position of AB for which the points O, C, and Q are in one straight line.

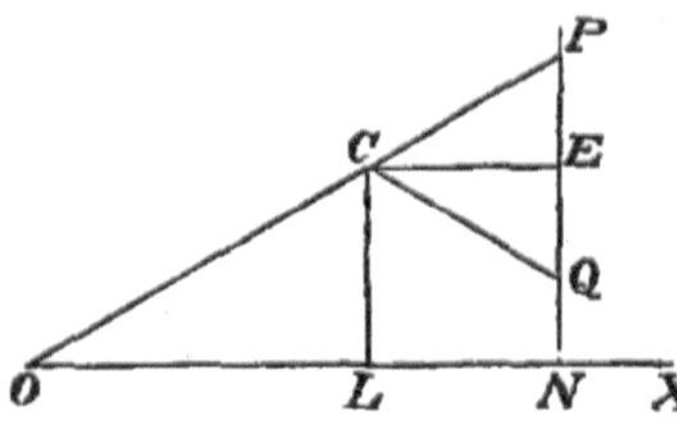

Let OX be this straight line, and let OC, CQ, be any other corresponding positions of the lines;
then, if CE is parallel to OX, CE bisects the angle PCQ, and, drawing PQN and CL perpendicular to OX,

$$QN = CL - PE, \quad PN = CL + PE,$$

hence
$$QN : PN :: OC - CP : OC + CP$$
$$:: OC - CQ : OC + CQ,$$

and $\therefore$ the locus of Q is an ellipse of which the semi-axes are $OC + CQ$ and $OC - CQ$.

If the straight lines through A and B perpendicular to OA and OB meet in K, the point K is the instantaneous centre of rotation. The normal to the path of Q is therefore QK and the tangent is the straight line through Q perpendicular to QK.

197. *Elliptic Compasses.* If two fine grooves, at right angles to each other, be made on the plane surface of a plate of wood or metal, and if two pegs, fastened to a straight rod, be made to move in these grooves, then a pencil attached to any point of the rod will trace out an ellipse.

By fixing the pencil at different points of the rod, we can obtain ellipses of any eccentricity, but of dimensions limited by the lengths of the rod and the grooves.

Burstow's Elliptograph.

OE is a groove in a stand which can be fixed to the paper or drawing board, and OA, OB are rods jointed at A, so that the end B can slip along the groove, while AO turns round the fixed end O.

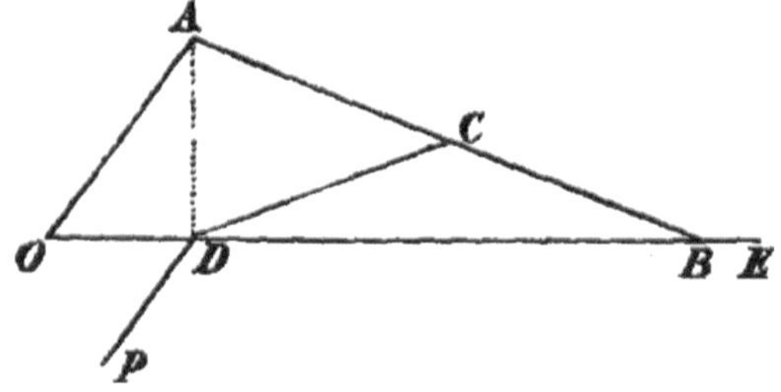

C is the middle point of AB, CD is a rod, the length of which is half that of AB, and the end D can slide along the groove.

It follows that the angle ADB is always a right angle.

A rod DP is taken of any convenient length, and, by means of a chain round the triangle ADC, is made to move so as to be always parallel to OA.

If the end B be moved along the groove, the end P will trace out an ellipse of which O is the centre, and the lengths of its semi-axes will be the length of DP and of the difference between the lengths of OA and DP. This can be seen by drawing a line OF perpendicular to OE, and producing DP to meet it in F. The motion will be that of a rod of length OA sliding between OE and OF. See Dyck, *Katalog der mathematischen Instrumente*, München, 1892.

MISCELLANEOUS PROBLEMS. I.

1. On a plane field the crack of the rifle and the thud of the ball striking the target are heard at the same instant; find the locus of the hearer.

2. PQ, $P'Q'$ are two focal chords of a parabola, and PR, parallel to $P'Q'$, meets in R the diameter through Q; prove that

$$PQ \,.\, P'Q' = PR^2.$$

3. CP and CD are conjugate semi-diameters of an ellipse; PQ is a chord parallel to one of the axes; shew that DQ is parallel to one of the straight lines which join the ends of the axes.

4. A line cuts two concentric, similar and similarly situated ellipses in P, Q, q, p. If the line move parallel to itself, $PQ \,.\, Qp$ is constant.

5. The portion of a tangent to an hyperbola intersected between the asymptotes subtends a constant angle at the focus.

6. If a circle be described passing through any point P of a given hyperbola and the extremities of the transverse axis, and the ordinate NP be produced to meet the circle in Q, the locus of Q is an hyperbola.

7. PQ is one of a series of chords inclined at a constant angle to the diameter AB of a circle; find the locus of the intersection of AP, BQ.

8. If from a point T in the director circle of an ellipse tangents TP, TP' be drawn, the line joining T with the intersection of the normals at P and P' passes through the centre.

9. The points, in which the tangents at the extremities of the transverse axis of an ellipse are cut by the tangent at any point of the curve, are joined, one with each focus; prove that the point of intersection of the joining lines lies in the normal at the point.

10. Having given a focus, the eccentricity, a point of the curve, and the tangent at the point, shew that in general two conics can be described.

11. A parabola is described with its focus at one focus of a given central conic, and touches the conic; prove that its directrix will touch a fixed circle.

12. The extremities of the latera recta of all conics which have a common transverse axis lie on two parabolas.

13. The tangent at a moveable point P of a conic intersects a fixed tangent in Q, and from S a straight line is drawn perpendicular to SQ and meeting in R the tangent at P; prove that the locus of R is a straight line.

14. On all parallel chords of a circle a series of isosceles triangles are described, having the same vertical angle, and having their planes perpendicular to the plane of the circle. Find the locus of their vertices; and find what the vertical angle must be in order that the locus may be a circle.

15. A series of similar ellipses whose major axes are in the same straight line pass through two given points. Prove that the major axes subtend right angles at four fixed points.

16. From the centre of two concentric circles a straight line is drawn to cut them in P and Q; through P and Q straight lines are drawn parallel to two given lines at right angles to each other. Shew that the locus of their point of intersection is an ellipse.

17. A circle always passes through a fixed point, and cuts a given straight line at a constant angle, prove that the locus of its centre is an hyperbola.

18. The area of the triangle formed by three tangents to a parabola is equal to one half that of the triangle formed by joining the points of contact.

19. If a parabola be described with any point on an hyperbola for focus and passing through the foci of the hyperbola, shew that its axis will be parallel to one of the asymptotes.

20. S and H being the foci, P a point in the ellipse, if HP be bisected in L, and AL be drawn from the vertex cutting SP in Q, the locus of Q is an ellipse whose focus is S.

21. If the diagonals of a quadrilateral circumscribing an ellipse meet in the centre the quadrilateral is a parallelogram.

22. A series of ellipses pass through the same point, and have a common focus, and their major axes of the same length; prove that the locus of their centres is a circle. What are the limits of the eccentricities of the ellipses, and what does the ellipse become at the higher limit?

23. If S, H be the foci of an hyperbola, LL' any tangent intercepted between the asymptotes, $SL \,.\, HL = CL \,.\, LL'$.

24. Tangents are drawn to an ellipse from a point on a similar and similarly situated concentric ellipse; shew that if P, Q be the points of contact, A, A' the ends of the axis of the first ellipse, the loci of the intersections of AP, $A'Q$, and of AQ, $A'P$ are two ellipses similar to the given ellipses.

25. Draw a parabola which shall touch four given straight lines. Under what condition is it possible to describe a parabola touching five given straight lines?

26. A fixed hyperbola is touched by a concentric ellipse. If the curvatures at the point of contact are equal the area of the ellipse is constant.

27. A circle passes through a fixed point, and cuts off equal chords AB, CD from two given parallel straight lines; prove that the envelope of each of the chords AD, BC is a central conic having the fixed point for one focus.

28. A straight line is drawn through the focus parallel to one asymptote and meeting the other; prove that the part intercepted between the curve and the asymptote is one-fourth the transverse axis, and the part between the curve and the focus one-fourth the latus rectum.

29. PQ is any chord of a parabola, cutting the axis in L; R, R' are the two points in the parabola at which this chord subtends a right angle: if RR' be joined, meeting the axis in L', LL' will be equal to the latus rectum.

30. If two equal parabolas have the same focus, tangents at points angularly equidistant from the vertices meet on the common tangent.

31. A parabola has its focus at S, and PSQ is any focal chord, while PP', QQ' are two chords drawn at right angles to PSQ at its extremities; shew that the focal chord drawn parallel to PP' is a mean proportional between PP' and QQ'.

32. With the orthocentre of a triangle as centre are described two ellipses, one circumscribing the triangle and the other touching its sides; prove that these ellipses are similar, and their homologous axes at right angles.

33. $ABCD$ is a quadrilateral, the angles at A and C being equal; a conic is described about $ABCD$ so as to touch the circumscribing circle of ABC at the point B; shew that BD is a diameter of the conic.

34. The volume of a cone cut off by a plane bears a constant ratio to the cube, the edge of which is equal to the minor axis of the section.

35. A tangent to an ellipse at P meets the minor axis in t, and tQ is perpendicular to SP; prove that SQ is of constant length, and that if PM be the perpendicular on the minor axis, QM will meet the major axis in a fixed point.

36. Describe an ellipse with a given focus touching three given straight lines, no two of which are parallel and on the same side of the focus.

37. Prove that the conic which touches the sides of a triangle, and has its centre at the centre of the nine-point circle, has one focus at the orthocentre, and the other at the centre of the circumscribing circle.

38. From Q, the middle point of a chord PP' of an ellipse whose focus is S, QG is drawn perpendicular to PP' to meet the major axis in G; prove that

$$2 \,.\, SG : SP + SP' :: SA : AX.$$

39. A straight rod moves in any manner in a plane; prove that, at any instant, the directions of motion of all its particles are tangents to a parabola.

40. If from a point T on the auxiliary circle, two tangents be drawn to an ellipse touching it in P and Q, and when produced meeting the circle again in p, q; shew that the angles PSp and QSq are together equal to the supplement of PTQ.

41. Tangents at the extremities of a pair of conjugate diameters of an ellipse meet in T; prove that ST, $S'T$ meet the conjugate diameters in four concyclic points.

42. From the point of intersection of an asymptote and a directrix of an hyperbola a tangent is drawn to the curve; prove that the line joining the point of contact with the focus is parallel to the asymptote.

43. If a string longer than the circumference of an ellipse be always drawn tight by a pencil, the straight portions being tangents to the ellipse, the pencil will trace out a confocal ellipse.

44. D is any point in a rectangular hyperbola from which chords are drawn at right angles to each other to meet the curve. If P, Q be the middle points of these chords, prove that P, Q, D and the centre of the hyperbola are concyclic.

45. From a point T in the auxiliary circle tangents are drawn to an ellipse, touching it in P and Q, and meeting the auxiliary circle again in p and q; shew that the angle pCq is equal to the sum of the angles PSQ and $PS'Q$.

46. The angle between the focal distance and tangent at any point of an ellipse is half the angle subtended at the focus by the diameter through the point.

47. H is a fixed point on the bisector of the exterior angle A of the triangle ABC; a circle is described upon HA as chord cutting the lines AB, AC in P and Q; prove that PQ envelopes a parabola which has H for focus, and for tangent at the vertex the straight line joining the feet of the perpendiculars from H on AB and AC.

48. Tangents to an ellipse, foci S and H, at the ends of a focal chord PHP' meet the further directrix in Q, Q'. The parabola, whose focus is S, and directrix PP', touches PQ, $P'Q'$, in Q, Q'; it also touches the normals at P, P', and the minor axis, and has for the tangent at its vertex the diameter parallel to PP'.

49. S is a fixed point, and E a point moving on the arc of a given circle; prove that the envelope of the straight line through E at right angles to SE is a conic.

50. A circle passing through a fixed point S cuts a fixed circle in P, and has its centre at O; the lines which bisect the angle SOP all touch a conic of which S is a focus.

51. The tangent to an ellipse at P meets the directrix, corresponding to S, in Z: through Z a straight line ZQR is drawn cutting the ellipse in Q, R; and the tangents at Q, R intersect (on SP) in T. Shew that a conic can be described with focus S, and directrix PZ, to pass through Q, R and T; and that TZ will be the tangent at T.

52. TP, TQ are tangents to an ellipse at P and Q; one circle touches TP at P and meets TQ in Q and Q'; another touches TQ at Q and meets TP in P and P'; prove that PQ' and QP' are divided in the same ratio by the ellipse.

53. If a chord $RPQV$ meet the directrices of an ellipse in R and V, and the circumference in P and Q, then RP and QV subtend, each at the focus nearer to it, angles of which the sum is equal to the angle between the tangents at P and Q.

54. Two tangents are drawn to the same branch of a rectangular hyperbola from an external point; prove that the angles which these tangents subtend at the centre are respectively equal to the angles which they make with the chord of contact.

55. If the normal at a point P of an hyperbola meet the minor axis in g, Pg will be to Sg in a constant ratio.

56. An ordinate NP of an ellipse is produced to meet the auxiliary circle in Q, and normals to the ellipse and circle at P and Q meet in R; RK, RL are drawn perpendicular to the axes; prove that KPL is a straight line, and also that $KP = BC$ and $LP = AC$.

57. If the tangent at any point P cut the axes of a conic, produced if necessary, in T and T', and if C be the centre of the curve, prove that the area of the triangle TCT' varies inversely as the area of the triangle PCN, where PN is the ordinate of P.

58. The circle of curvature of an ellipse at P passes through the focus S, SM is drawn parallel to the tangent at P to meet the diameter PCP' in M; shew that it divides this diameter in the ratio of $3:1$.

59. Prove the following construction for a pair of tangents from any external point T to an ellipse of which the centre is C: join CT, let $TPCP'T$ a similar and similarly situated ellipse be drawn, of which CT is a diameter, and P, P' are its points of intersection with the given ellipse; TP, TP' will be tangents to the given ellipse.

60. Through a fixed point a pair of chords of a circle are drawn at right angles: prove that each side of the quadrilateral formed by joining their extremities envelopes a conic of which the fixed point and the centre of the circle are foci.

61. Any conic passing through the four points of intersection of two rectangular hyperbolas will be itself a rectangular hyperbola.

62. R is the middle point of a chord PQ of a rectangular hyperbola whose centre is C. Through R, RQ', RP' are drawn parallel to the tangents at P and Q respectively, meeting CQ, CP in Q', P'. Prove that C, P', R, Q' are concyclic.

63. The tangents at two points Q, Q' of a parabola meet the tangent at P in R, R' respectively, and the diameter through their point of intersection T meets it in K; prove that $PR = KR'$, and that, if QM, $Q'M'$, TN be the ordinates of Q, Q', T respectively to the diameter through P, PN is a mean proportional between PM and PM'.

64. Common tangents are drawn to two parabolas, which have a common directrix, and intersect in P, Q: prove that the chords joining the points of contact in each parabola are parallel to PQ, and the part of each tangent between its points of contact with the two curves is bisected by PQ produced.

65. An ellipse has its centre on a given hyperbola and touches the asymptotes. The area of the ellipse being always a maximum, prove that its chord of contact with the asymptotes always touches a similar hyperbola.

66. A circle and parabola have the same vertex A and a common axis. $BA'C$ is the double ordinate of the parabola which touches the circle at A', the other

extremity of the diameter which passes through A; PP' is any other ordinate of the parabola parallel to this, meeting the axis in N and the chord AB produced in R: shew that the rectangle between RP and RP' is proportional to the square on the tangent drawn from N to the circle.

67. Tangents are drawn at two points, P, P' on an ellipse. If any tangent be drawn meeting those at P, P' in R, R', shew that the line bisecting the angle RSR' intersects RR' on a fixed tangent to the ellipse. Find the point of contact of this tangent.

68. Having given a pair of conjugate diameters of an ellipse, PCP', DCD', let PF be the perpendicular from P on CD, in PF take PE equal to CD, bisect CE in O, and on CE as diameter describe a circle; prove that PO will meet the circle in two points Q and R such that CQ, CR are the directions of the semi-axes, and PQ, PR their lengths.

69. A straight line is drawn through the angular point A of a triangle ABC to meet the opposite side in a; two points O, O' are taken on Aa, and CO, CO' meet AB in c and c', and BO, BO' meet CA in b, b'; shew that a conic passing through $abb'cc'$ will be touched by BC.

70. If TP, TQ are two tangents to a parabola, and any other tangent meets them in Q and R, the middle point of QR describes a straight line.

71. Lines from the centre to the points of contact of two parallel tangents to a rectangular hyperbola and concentric circle make equal angles with either axis of the hyperbola.

72. A line moves between two lines at right angles so as to subtend a right angle and a half at a fixed point on the bisector of the right angle; prove that it touches a rectangular hyperbola.

73. Two cones, whose vertical angles are supplementary, are placed with their vertices coincident and their axes at right angles, and are cut by a plane perpendicular to a common generating line; prove that the directrices of the section of one cone pass through the foci of the section of the other.

74. The normal at a point P of an ellipse meets the curve again in P', and through O, the centre of curvature at P, the chord QOQ' is drawn at right angles to PP'; prove that

$$QO \,.\, OQ' : PO \,.\, OP' :: 2 \,.\, PO : PP'.$$

75. From an external point T, tangents are drawn to an ellipse, the points of contact being on the same side of the major axis. If the focal distances of these points intersect in M and N, TM, TN are tangents to a confocal hyperbola, which passes through M and N.

76. Two tangents to an hyperbola from T meet the directrix in F and F'; prove that the circle, centre T, which touches SF, SF', meets the directrix in two points the radii to which from the point T are parallel to the asymptotes.

77. QR, touching the ellipse at P, is one side of the parallelogram formed by tangents at the ends of conjugate diameters; if the normal at P meet the axes in G and g, prove that QG and Rg are at right angles.

78. If PP' be a double ordinate of an ellipse, and if the normal at P meet CP' in O, prove that the locus of O is a similar ellipse, and that its axis is to the axis of the given ellipse in the ratio

$$AC^2 - BC^2 : AC^2 + BC^2.$$

79. A chord of a conic whose pole is T meets the directrices in R and R'; if SR and $S'R'$ meet in Q, prove that the minor axis bisects TQ.

80. On a parabola, whose focus is S, three points Q, P, Q' are taken such that the angles PSQ, PSQ' are equal; the tangent at P meets the tangents at Q, Q' in T, T': shew that

$$TQ : T'Q' :: SQ : SQ'.$$

81. If from any point P of a parabola perpendiculars PN, PL are let fall on the axis and the tangent at the vertex, the line LN always touches another parabola.

82. PQ is any diameter of a section of a cone whose vertex is V; prove that $VP + VQ$ is constant.

83. If SY, SK are the perpendiculars from a focus on the tangent and normal at any point of a conic, the straight line YK passes through the centre of the conic.

84. If the axes of two parabolas are in the same direction, their common chord bisects their common tangents.

85. Find the position of the normal chord which cuts off from a parabola the least segment.

86. From the point in which the tangent at any point P of an hyperbola meets either asymptote perpendiculars PM, PN are let fall upon the axes. Prove that MN passes through P.

87. If two parabolas whose latera recta have a constant ratio, and whose foci are two given points S, S', have a contact of the second order at P, the locus of P is a circle.

88. Find the class of plane curves such that, if from a fixed point in the plane, perpendiculars are let fall on the tangent and normal at any point of any one of the curves, the join of the feet of the perpendiculars will pass through another fixed point.

89. If two ellipses have one common focus S and equal major axes, and if one ellipse revolves in its own plane about S, the chord of intersection envelopes a conic confocal with the fixed ellipse.

90. The tangent at any point P of an ellipse meets the axis minor in T and the focal distances SP, HP meet it in R, r. Also ST, HT, produced if necessary, meet the normal at P in Q, q, respectively. Prove that Qr and qR are parallel to the axis major.

91. Two points describe the circumference of an ellipse, with velocities which are to one another in the ratio of the squares on the diameters parallel to their respective directions of motion. Prove that the locus of the point of intersection of their directions of motion will be an ellipse, confocal with the given one.

92. If AA' be the axis major of an elliptic section of a cone, vertex O, and if AG, $A'G'$ perpendicular to AV, $A'V$ meet the axis of the cone in G and G', and GU, $G'U'$ be the perpendiculars let fall on AA', prove that U and U' are the centres of curvature at A and A'.

93. By help of the geometry of the cone, or otherwise, prove that the sum of the tangents from any point of an ellipse to the circles of curvature at the vertices is constant.

94. If two tangents be drawn to a section of a cone, and from their intersection two straight lines be drawn to the points where the tangent plane to the cone through one of the tangents touches the focal spheres, prove that the angle contained by these lines is equal to the angle between the tangents.

95. If CP, CD are conjugate semi-diameters and if through C is drawn a line parallel to either focal distance of P, the perpendicular from D upon this line will be equal to half the minor axis.

96. The area of the parallelogram formed by the tangents at the ends of any pair of diameters of a central conic varies inversely as the area of the parallelogram formed by joining the points of contact.

97. Shew how to draw through a given point a plane which will have the given point for (1) focus, (2) centre, of the section it makes of a given right circular cone: noticing any limitations in the position of the point which may be necessary.

98. In the first figure of Art. 148, if a plane be drawn intersecting the focal spheres in two circles and the cone in an ellipse, the sum or difference of the tangents from any point of the ellipse to the circles is constant.

99. If sections of a right cone be made, perpendicular to a given plane, such that the distance between a focus of a section and that vertex which lies on one of the generating lines in the given plane be constant, prove that the transverse axes, produced if necessary, of all sections will touch one of two fixed circles.

100. A sphere rolls in contact with two intersecting straight wires; prove that its centre describes an ellipse.

CHAPTER XI.

HARMONIC PROPERTIES, POLES AND POLARS.

198. DEF. *A straight line is harmonically divided in two points when the whole line is to one of the extreme parts as the other extreme part is to the middle part.*

Thus AD is harmonically divided in C and B, when

$$AD : AC :: BD : BC.$$

A C B D

This definition may also be presented in the following form.

The straight line AB is harmonically divided in C and D, when it is divided internally in C, and externally in D, in the same ratio.

Under these circumstances the four points A, C, B, D constitute *an Harmonic Range*, and if through any point O four straight lines OA, OC, OB, OD be drawn, these four lines constitute an *Harmonic Pencil.*

PROP. I. *If a straight line be drawn parallel to one of the rays of an harmonic pencil, its segments made by the other three will be equal, and any straight line is divided harmonically by the four rays.*

Let $ACBD$ be the given harmonic range, and draw ECF through C parallel to OD, and meeting OA, OB in E and F.

Then $$AD : AC :: OD : EC,$$

and $$BD : BC :: OD : CF;$$

but from the definition

$$AD : AC :: BD : BC;$$

$$\therefore EC = CF,$$

and any other line parallel to ECF is obviously bisected by OC.

Next, let $acbd$ be any straight line cutting the pencil, and draw ecf parallel to Od; so that $ec = cf$.

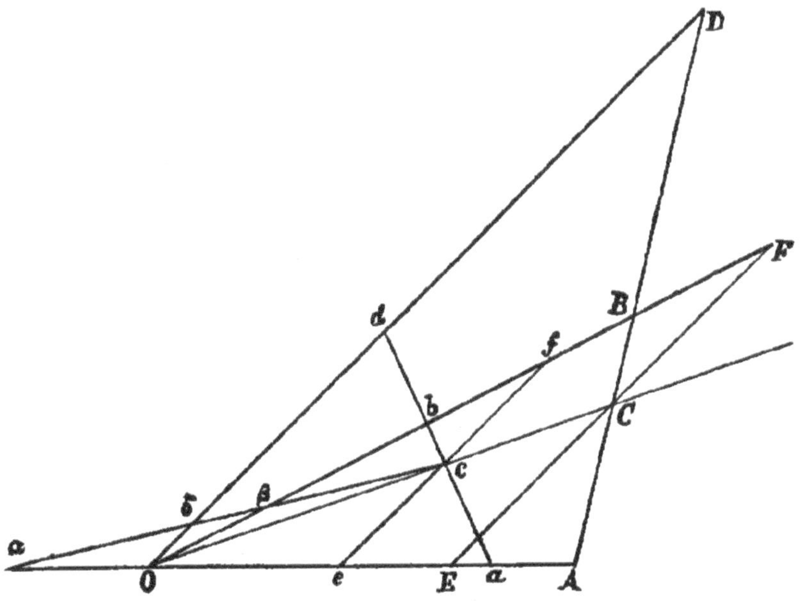

Then $\quad ad : ac :: Od : ec,$

and $\quad bd : bc :: Od : cf;$

$$\therefore ad : ac :: bd : bc;$$

that is, *acbd* is harmonically divided.

If the line $c\beta\delta\alpha$ be drawn cutting AO produced,

then $\quad \alpha\delta : \alpha c :: O\delta : ec,$

and $\quad \beta\delta : \beta c :: O\delta : cf;$

$$\therefore \alpha\delta : \alpha c :: \beta\delta : \beta c,$$

or $\quad \alpha c : \alpha\delta :: \beta c : \beta\delta,$

and similarly it may be shewn in all other cases that the line is harmonically divided.

199. PROP. II. *The pencil formed by two straight lines and the bisectors of the angles between them is an harmonic pencil.*

For, if OA, OB be the lines, and OC, OD the bisectors, draw KPL parallel to OC and meeting OA, OD, OB. Then the angles OKL, OLK are obviously equal, and the angles at P are right angles; therefore $KP = PL$, and the pencil is harmonic.

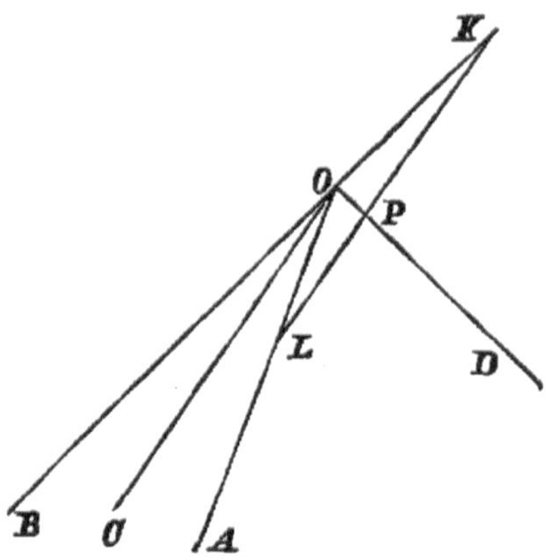

200. Prop. III. *If $ACBD$, $Acbd$ be harmonic ranges, the straight lines Cc, Bb, Dd will meet in a point, as also Cd, cD, Bb.*

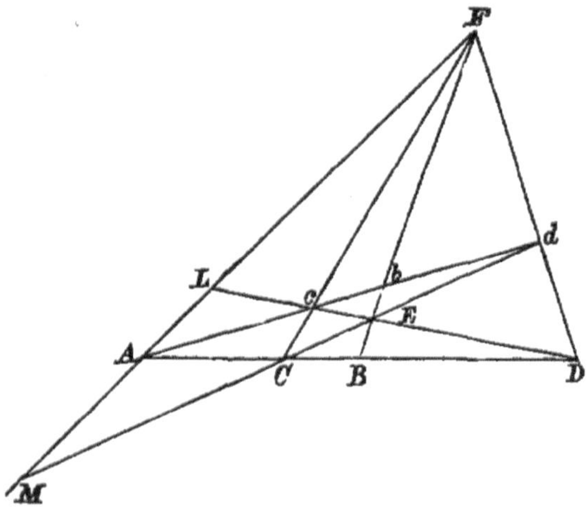

For, if Cc, Dd meet in F, join Fb; then the pencil $F(Acbd)$ is harmonic, and will be cut harmonically by AD.

Hence Fb produced will pass through B.

Similarly, if Cd, cD meet in E,
$E(Acbd)$ is harmonic, and therefore bE produced will pass through B.

Harmonic Properties of a Quadrilateral.

In the preceding figure, let $CcdD$ be any quadrilateral; and let dc, DC meet in A, Cd, cD in E, and Cc, Dd in F.

Then taking b and B so as to divide Acd and ACD harmonically, the ranges $Acbd$ and $ACBD$ are harmonic, and therefore Bb passes through both E and F.

Similarly it can be shewn that AF is divided harmonically in L and M, by Dc and dC.

For $E(Acbd)$ is harmonic and therefore the transversal $ALFM$ is harmonically divided.

201. PROP. IV. *If $ACBD$ be an harmonic range, and E the middle point of CD,*

$$EA \,.\, EB = EC^2.$$

A C B E D

For
$$AD : AC :: BD : BC,$$
or
$$AE + EC : AE - EC :: EC + EB : EC - EB;$$
$$\therefore AE : EC :: EC : EB,$$
or
$$AE \,.\, EB = EC^2 = ED^2.$$

Hence also, conversely, if $EC^2 = ED^2 = AE \,.\, EB$, the range $ACBD$ is harmonic, C and D being on opposite sides of E.

Hence, if a series of points A, a, B, b, ... on a straight line be such that

$$EA \,.\, Ea = EB \,.\, Eb = EC \,.\, Ec \ldots$$
$$= EP^2,$$

and if $EQ = EP$, then the several ranges $(APaQ)$, $(BPbQ)$, &c. are harmonic.

202. DEF. A system of pairs of points on a straight line such that

$$EA \,.\, Ea = EB \,.\, EB \,.\, Eb = \ldots \quad = EP^2 = EQ^2$$

is called a system in *Involution*, the point E being called the centre and P, Q the foci of the system.

Any two corresponding points A, a, are called *conjugate* points, and it appears from above that any two conjugate points form, with the foci of the system, an harmonic range.

It will be noticed that a focus is a point at which conjugate points coincide, and that the existence of a focus is only possible when the points A and a are both on the same side of the centre.

203. PROP. V. *Having given two pairs of points, A and a, B and b, it is required to find the centre and foci of the involution.*

If E be the centre,

$$EA : EB :: Eb : Ea;$$

$$\therefore EA : AB :: Eb : ab,$$

or

$$EA : Eb :: AB : ab.$$

This determines E, and the foci P and Q are given by the relations

$$EP^2 = EQ^2 = EA \,.\, EA.$$

We shall however find the following relation useful.

Since $$EA : Eb :: EB : Ea;$$

$$\therefore EA : Ab :: EB : aB,$$

or $$EA : EB :: Ab : aB;$$

but $$Eb : EA :: ab : AB;$$

$$\therefore Eb : EB :: Ab \,.\, ba : AB \,.\, Ba.$$

Again, $$Qb : Pb :: QB : PB;$$

$$\therefore Qb - Pb : Pb :: QB - PB : PB,$$

$$2 \,.\, EP : Pb :: 2 \,.\, EB : BP;$$

$$\therefore Pb^2 : PB^2 :: EP^2 : EB^2,$$

$$:: Eb : EB$$

$$:: Ab \,.\, ba : AB \,.\, Ba.$$

This determines the ratio in which Bb is divided by P.

204. If $QAPa$ be an harmonic range and E the middle point of PQ, and if a circle be described on PQ as diameter, the lines joining any point R on this circle with P and Q will bisect the angles between AR and aR.

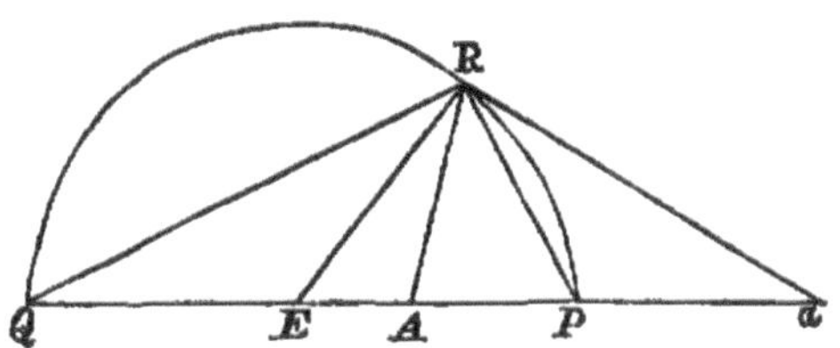

For $$EA \,.\, Ea = EP^2 = ER^2;$$
$$\therefore EA : ER :: ER : Ea,$$
and the triangles ARE, aRE are similar.

Hence $$AR : aR :: EA : ER$$
$$:: EA : EP.$$

But $$Ea : EP :: EP : EA;$$
$$\therefore aP : EP :: AP : EA.$$

Hence $$AR : aR :: AP : aP,$$
and ARa is bisected by RP.

Hence, if A and a, B and b be conjugate points of a system in involution of which P and Q are the foci, it follows that AB and ab subtend equal angles at any point of the circle on PQ as diameter.

This fact also affords a means of obtaining the relations of Art. 203.

We must observe that if the points A, a are on one side of the centre and B, b on the other, the angles subtended by AB, ab are supplementary to each other.

205. PROP. VI. *If four points form an harmonic range, their conjugates also form an harmonic range.*

Let A, B, C, D be the four points, a, b, c, d their conjugates.

Then, as in the eighth line of Art. 203,
$$EA : Ed :: AD : ad,$$
or $$ED : Ea :: AD : ad;$$
$$\therefore AD \,.\, Ea = ED \,.\, ad.$$

Similarly $$AC \,.\, Ea = EC \,.\, ac,$$
$$BD \,.\, Eb = ED \,.\, bd,$$
$$BC \,.\, Eb = EC \,.\, bc.$$

But, $ABCD$ being harmonic,
$$AD : AC :: BD : BC;$$
$$\therefore ED \,.\, ad : EC \,.\, ac :: ED \,.\, bd : EC \,.\, bc.$$

Hence $$ad : ac :: bd : bc,$$
or the range of the conjugates is harmonic.

206. PROP. VII. *If a system of conics pass through four given points, any straight line will be cut by the system in a series of points in involution.*

The four fixed points being C, D, E, F, let the line meet one of the conics in A and a, and the straight lines CF, ED, in B and b.

Then the rectangles $AB \,.\, Ba$, $CB \,.\, BF$ are in the ratio of the squares on parallel diameters, as also are $Ab \,.\, ba$ and $Db \,.\, bE$.

But the squares on the diameters parallel to CF, ED are in the constant ratio $KF \,.\, KC : KE \,.\, KD$; and, the line Bb being given in position, the rectangles $CB \,.\, BF$ and $Db \,.\, bE$ are given; therefore the rectangles $AB \,.\, Ba$, $Ab \,.\, ba$ are in a constant ratio.

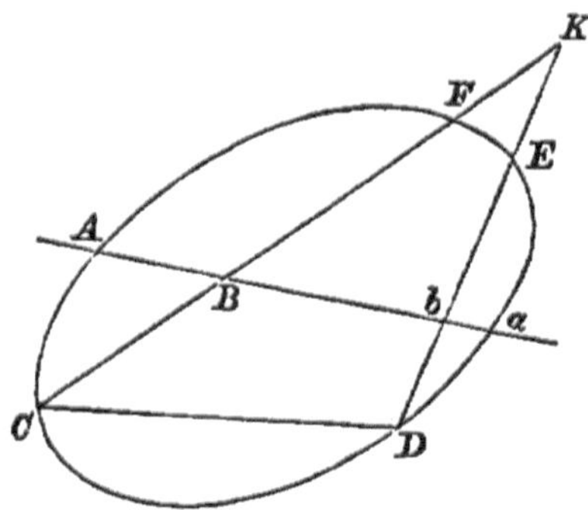

But (Art. 203) this ratio is the same as that of PB^2 to Pb^2, if P be a focus of the involution A, a, B, b.

Hence P is determined, and all the conics cut the line Bb in points which form with B, b a system in involution.

We may observe that the foci are the points of contact of the two conics which can be drawn through the four points touching the line, and that the centre is the intersection of the line with the conic which has one of its asymptotes parallel to the line.

207. PROP. VIII. *If through any point two tangents be drawn to a conic, any other straight line through the point will be divided harmonically by the curve and the chord of contact.*

Let AB, AC be the tangents, $ADFE$ the straight line.

Through D and E draw $GDHK$, $LEMN$ parallel to BC.

Then the diameter through A bisects DH, and BC, and therefore bisects GK; hence $GD = HK$, and similarly $LE = MN$.

Also $$LE : EN :: GD : DK;$$
$$\therefore LE . EN : LE^2 :: GD . DK : GD^2,$$
or $$LE . LM : GD . GH :: LE^2 : GD^2$$
$$:: LA^2 : GA^2.$$

But $$LE . LM : GD . GH :: LB^2 : BG^2;$$
hence $$AL : AG :: BL : BG,$$

and therefore $$AE : AD :: FE : FD,$$
that is, $ADFE$ is harmonically divided.

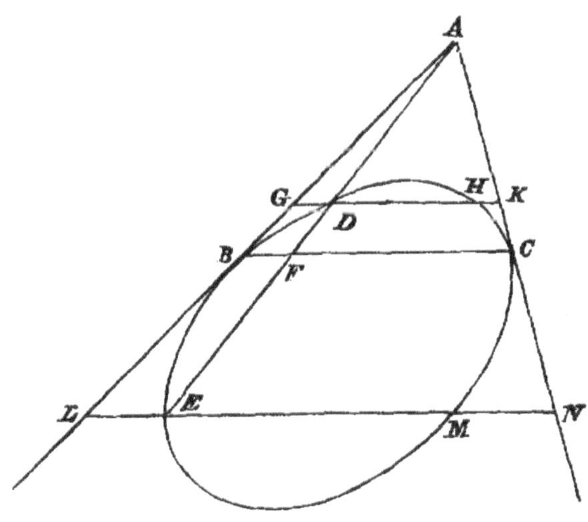

208. Prop. IX. *If two tangents be drawn to a conic, any third tangent is harmonically divided by the two tangents, the curve, and the chord of contact.*

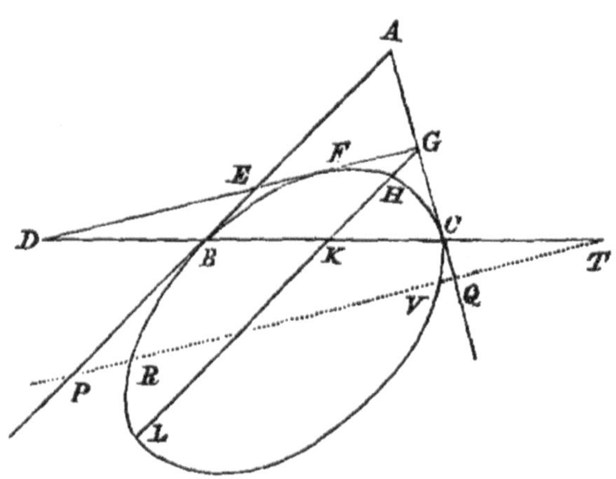

Let $DEFG$ be the third tangent, and through G, the point in which it meets AC, draw $GHKL$ parallel to AB, cutting the curve and the chord of contact in H, K, L.

Then $$GH\,.\,GL : GC^2 :: AB^2 : AC^2$$
$$:: GK^2 : GC^2;$$
$$\therefore GH\,.\,GL = GK^2.$$

Hence $$DG^2 : DE^2 :: GK^2 : EB^2$$
$$:: GH\,.\,GL : EB^2$$
$$:: FG^2 : FE^2;$$

that is, $DEFG$ is an harmonic range.

209. Prop. X. *If any straight line meet two tangents to a conic in P and Q, the chord of contact in T and the conic in R and V,*

$$PR\,.\,PV : QR\,.\,QV :: PT^2 : QT^2.$$

Taking the preceding figure, draw the tangent $DEFG$ parallel to PQ.

Then $$PR\,.\,PV : EF^2 :: PB^2 : BE^2$$
$$:: PT^2 : DE^2;$$
and $$QR\,.\,QV : GF^2 :: QC^2 : GC^2$$
$$:: QT^2 : DG^2;$$
but $$EF : DE :: GF : DG;$$
$$\therefore PR\,.\,PV : PT^2 :: QR\,.\,QV : QT^2.$$

210. PROP. XI. *If chords of a conic be drawn through a fixed point the pairs of tangents at their extremities will intersect in a fixed line.*

Let B be the fixed point and C the centre, and let CB meet the curve in P.

Take A in CP such that
$$CA : CP :: CP : CB;$$
then B is the middle point of the chord of contact of the tangents AQ, AR.

Draw any chord EBF, and let the tangents at E and F meet in G: also join CG and draw PN parallel to EF.

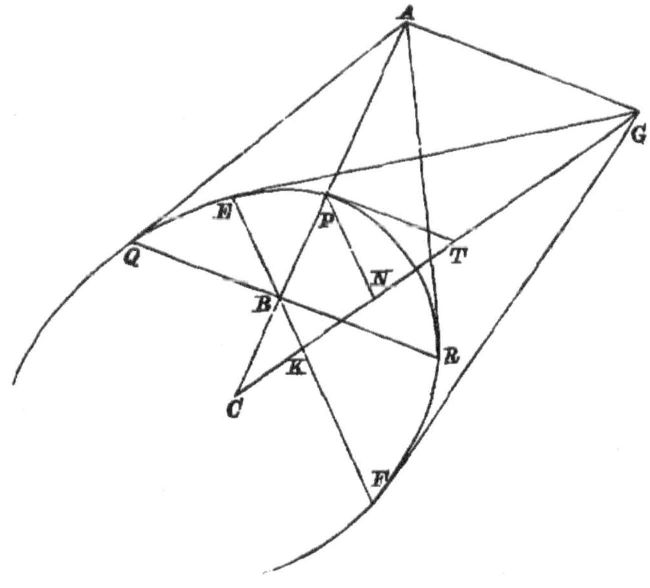

Then if CG meet EF in K and the tangent at P in T,

$$CK \,.\, CG = CN \,.\, CT;$$

$$\therefore CG : CT :: CN : CK$$

$$:: CP : CB$$

$$:: CA : CP;$$

hence AG is parallel to PT, and the point G therefore lies on a fixed line.

If the conic be a parabola, we must take AP equal to BP: then, remembering that KG and NT are bisected by the curve, the proof is the same as before.

211. If A be the fixed point, let CA meet the curve in P, and take B in CP such that

$$CB : CP :: CP : CA;$$

then B is the middle point of the chord of contact of the tangents AQ, AR. Draw any chord AEF, and let the tangents at E and F meet in G; also join CG and draw PN parallel to EF.

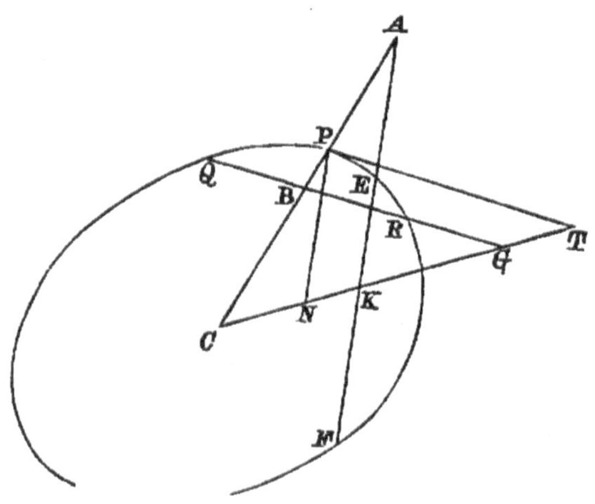

Then

$$CK \,.\, CG = CN \,.\, CT;$$

$$\therefore CG : CT :: CN : CK$$

$$:: CP : CA$$

$$:: CB : CP;$$

$\therefore BG$ is parallel to PT and coincides with the chord of contact QR.

Hence, conversely, if from points on a straight line pairs of tangents be drawn to a conic, the chords of contact will pass through a fixed point.

Poles and Polars.

212. DEF. The straight line which is the locus of the points of intersection of tangents at the extremities of chords through a fixed point is called the *polar* of the point.

Also, if from points in a straight line pairs of tangents be drawn to a conic, the point in which all the chords of contact intersect is called the *pole* of the line.

If the pole be without the curve the polar is the chord of contact of tangents from the pole.

If the pole be on the curve the polar is the tangent at the point.

It follows at once from these definitions that the focus of a conic is the pole of the directrix, and that the foot of the directrix is the pole of the latus rectum.

213. PROP. XII. *A straight line drawn through any point is divided harmonically by the point, the curve, and the polar of the point.*

If the point be without the conic this is already proved in Art. 207.

If it be within the conic, as B in the figure of Art. 210, then, drawing any chord $FBEV$ meeting in V the polar of B, which is AG, the chord of contact of tangents from V passes through B, by Art. 211, and the line $VEBF$ is therefore harmonically divided.

Hence the polar may be constructed by drawing two chords through the pole and dividing them harmonically; the line joining the points of division is the polar.

Or, in the figure of Art. 210,

$$CB \, . \, CA = CP^2,$$

so that the polar of B is obtained by taking the point A on the diameter through B, at the distance from C given by the above relation, and then drawing AG parallel to the diameter which is conjugate to CP.

COR. Hence it follows that *the centre of a conic is the pole of a line at an infinite distance.*

For, if CB is diminished indefinitely, CA is increased indefinitely.

214. PROP. XIII. *The polars of two points intersect in the pole of the line joining the two points.*

For, if A, B be the two points and O the pole of AB, the line AO is divided harmonically by the curve, and therefore the polar of A passes through the point O.

Similarly the polar of B passes through O;

That is, the polars of A and B intersect in the pole of AB.

215. PROP. XIV. *If a quadrilateral be inscribed in a conic, its opposite sides and diagonals will intersect in three points such that each is the pole of the line joining the other two.*

Let $ABCD$ be the quadrilateral, F and G the points of intersection of AD, BC, and of DC, AB.

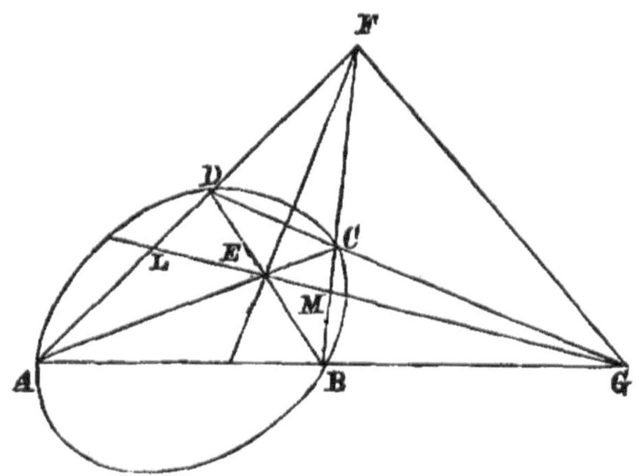

Let EG meet FA, FB, in L and M.

Then (Art. 200) $FDLA$ and $FCMB$ are harmonic ranges;

Therefore L and M are both on the polar of F (Art. 213), and EG is the polar of F.

Similarly, EF is the polar of G, and therefore E is the pole of FG (Art. 214).

216. DEF. If each of the sides of a triangle be the polar, with regard to a conic, of the opposite angular point, the triangle is said to be *self-conjugate* with regard to the conic.

Thus the triangle EGF in the above figure is self-conjugate.

To construct a self-conjugate triangle, take a straight line AB and find its pole C.

Draw through C any straight line CD cutting AB in D, and find the pole E of CD, which lies on AB: then CDE is self-conjugate.

217. PROP. XV. *If a quadrilateral circumscribe a conic, its three diagonals form a self-conjugate triangle.*

Let the polar of F (that is, the chord of contact $P'P$), meet FG in R; then, since R is on the polar of F, it follows that F is on the polar of R.

Now $F(AEBG)$ is harmonic (Art. 200), and, if FE meet $P'P$ in T, $P'TPR$ is an harmonic range; hence, by the theorem of Art. 213, FT, *i.e.* FE, is the polar of R.

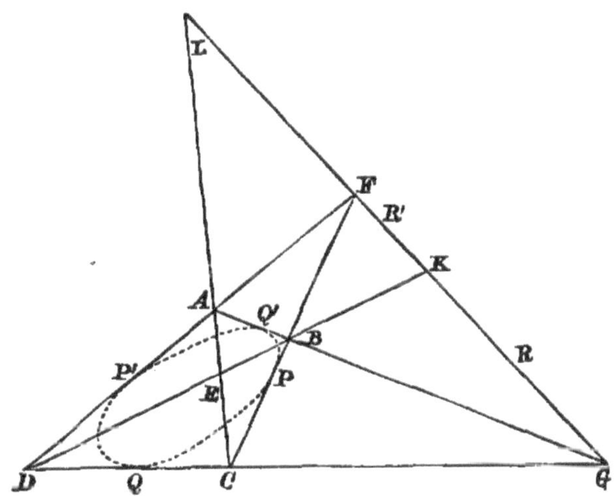

Similarly, if the other chord of contact QQ' meet FG in R', GE is the polar of R';

$$\therefore E \text{ is the pole of } RR', \text{ that is, of } LK.$$

Again, $DEBK$ is harmonic, and therefore the pencil $C(QEPK)$ is harmonic.

Hence, if QP meet AC in S and CK in V, $QSPV$ is harmonic, and therefore S is on the polar of V.

But S is on the polar of C; therefore CV, that is, CK, is the polar of S.

Similarly, if $P'Q'$ meet AC in S', AK is the polar of S'.

Hence it follows that K is the pole of SS', that is, of EL; ELK is therefore a self-conjugate triangle.

218. PROP. XVI. *If a system of conics have a common self-conjugate triangle, any straight line passing through one of the angular points of the triangle is cut in a series of points in involution.*

For, if ABC be the triangle, and a line $APDQ$ meet BC in D, and the conic in P and Q, $APDQ$ is an harmonic range, and all the pairs of points P, Q form with A and D an harmonic range.

Hence the pairs of points form a system in involution, of which A and D are the foci.

219. PROP. XVII. *The pencil formed by the polars of the four points of an harmonic range is an harmonic pencil.*

Let $ABCD$ be the range, O the pole of AD.

Let the polars Oa, Ob, Oc, Od meet AD in a, b, c, d, and let AD meet the conic in P and Q.

Then $APaQ$, $CPcQ$, &c. are harmonic ranges; and therefore (Arts. 201, 202) a, c, b, d are the conjugates of A, C, B, D.

Hence (Art. 205) the range $acbd$ is harmonic, and therefore the pencil O $(acbd)$ is harmonic.

EXAMPLES.

1. If PSP' is a focal chord of a conic, any other chord through S is divided harmonically by the directrix and the tangents at P and P'.

2. If two sections of a right cone be taken, having the same directrix, the straight line joining the corresponding foci will pass through the vertex.

3. If a series of circles pass through the same two points, any transversal will be cut by the circles in a series of points in involution.

4. If O be the centre of the circle circumscribing a triangle ABC, and $B'C'$, $C'A'$, $A'B'$, the respective polars with regard to a concentric circle of the points A, B, C, prove that O is the centre of the circle inscribed in the triangle $A'B'C'$.

5. OA, OB, OC being three straight lines given in position, shew that there are three other straight lines each of which forms with OA, OB, OC an harmonic pencil; and that each of the three OA, OB, OC forms with the second three an harmonic pencil.

6. The straight line $ACBD$ is divided harmonically in the points C, B; prove that if a circle be described on CD as diameter, any circle passing through A and B will cut it at right angles.

7. Three straight lines AD, AE, AF are drawn through a fixed point A, and fixed points C, B, D are taken in AD, such that $ACBD$ is an harmonic range. Any straight line through B intersects AE and AF in E and F, and CE, DF intersect in P; DE, CF in Q. Shew that P and Q always lie in a straight line through A, forming with AD, AE, AF an harmonic pencil.

8. CA, CB are two tangents to a conic section, O a fixed point in AB, POQ any chord of the conic; prove that the intersections of AP, BQ, and also of AQ, BP lie in a fixed straight line which forms with CA, CO, CB an harmonic pencil.

9. If three conics pass through the same four points, the common tangent to two of them is divided harmonically by the third.

10. Two conics intersect in four points, and through the intersection of two of their common chords a tangent is drawn to one of them; prove that it is divided harmonically by the other.

11. Prove that the two tangents through any point to a conic, any line through the point and the line to the pole of the last line, form an harmonic pencil.

12. The locus of the poles, with regard to the auxiliary circle, of the tangents to an ellipse, is a similar ellipse.

13. The asymptotes of an hyperbola and any pair of conjugate diameters form an harmonic pencil.

14. PSQ and $PS'R$ are two focal chords of an ellipse; two other ellipses are described having P for a common focus, and touching the first ellipse at Q and R respectively. The three ellipses have equal major axes. Prove that the directrices of the last two ellipses pass through the pole of QR.

15. Tangents from T touch an ellipse in P and Q, and PQ meets the directrices in R and R'; shew that PR and QR' subtend equal angles at T.

16. The poles of a given straight line, with respect to sections through it of a given cone, all lie upon a straight line passing through the vertex of the cone.

17. If from a given point in the axis of a conic a chord be drawn, the perpendicular from the pole of the chord upon the chord will meet the axis in a fixed point.

18. Q is any point in the tangent at a point P of a conic; QG perpendicular to CP meets the normal at P in G, and QE perpendicular to the polar of Q meets the normal at P in E; prove that EG is constant and equal to the radius of curvature at P.

19. The line joining two fixed points A and B meets the two fixed lines OP, OQ in P and Q.

A conic is described so that OP and OQ are the polars of A and B with respect to it. Prove that the locus of its centre is the line OR, where R divides AB so that

$$AR : RB :: QR : RP.$$

20. If from a point O in the normal at a point R of an ellipse tangents OP, OQ are drawn, the angles PRO, QRO are equal.

21. The focal distances of a point on a conic meet the curve again in Q, R; shew that the pole of QR will lie upon the normal at the first point.

22. The tangent at any point A of a conic is cut by two other tangents and their chord of contact in B, C, D; shew that $(ABDC)$ is harmonic.

23. A rectangular hyperbola circumscribes a triangle ABC; if D, E, F be the feet of the perpendiculars from A, B, C on the opposite sides, the loci of the poles of the sides of the triangle ABC are the lines EF, FD, DE.

24. Two common chords of a given ellipse and a circle pass through a given point; shew that the locus of the centres of all such circles is a straight line through the given point.

25. If $ABCD$ is a quadrilateral inscribed in a conic, and if AD, BC meet in P, and AC, BD in Q, PQ passes through the pole of AB.

26. PCP' is any diameter of an ellipse. The tangents at the points D, E intersect in F, and PE, $P'D$ intersect in G. Shew that FG is parallel to DCD'.

27. PP' is a chord of a conic, QQ' any chord through its pole. Prove that lines drawn from P parallel to the tangents at Q and Q' to meet $P'Q$, and $P'Q'$ respectively are bisected by QQ'.

28. If the pencil joining four fixed points on a conic to any one point on the conic is harmonic, the pencil joining the fixed points to any point on the conic is harmonic.

29. If PQ is the chord of a conic having its pole on the chord AB or AB produced, and if Qq is the chord parallel to AB, then Pq bisects AB.

30. If a quadrilateral circumscribe a conic, the intersection of the lines joining opposite points of contact is the same as the intersection of the diagonals.

CHAPTER XII.

Reciprocal Polars.

220. The pole of a line with regard to any conic being a point and the polar of a point a line, it follows that any system of points and lines can be transformed into a system of lines and points.

This process is called *reciprocation*, and it is clear that any theorem relating to the original system will have its analogue in the system formed by reciprocation.

Thus, if a series of lines be concurrent, the corresponding points are collinear; and the theorem of Art. 219 is an instance of the effect of reciprocation.

221. Def. If a point move in a curve (C), its polar will always touch some other curve (C'); this latter curve is called the reciprocal polar of (C) with regard to the auxiliary conic.

Prop. I. *If a curve C' be the polar of C, then will C be the polar of C'.*

For, if P, P' be two consecutive points of C, the intersection of the polars of P and P' is a point Q, which is the pole of the line PP'.

But the point Q is ultimately, when P and P' coincide, the point of contact of the curve which is touched by the polar of P.

Hence the polar of any point Q of C' is a tangent to the curve C.

222. So far we have considered poles and polars generally with regard to any conic; we shall now consider the case in which a circle is the auxiliary curve.

In this case, if AB be a line, P its pole, and CY the perpendicular from the centre of the circle on AB, the rectangle $CP.\ CY$ is equal to the square on the radius of the circle.

A simple construction is thus given for the pole of a line, or the polar of the point.

As an illustration take the theorem of the existence of the orthocentre in a triangle.

Let AOD, BOE, COF be the perpendiculars, O being the orthocentre.

The polar reciprocal of the line BC is a point A', and of the point A a line $B'C'$.

To the line AD corresponds a point P on $B'C'$, and since ADB is a right angle, it follows that PSA' is a right angle, S being the centre of the auxiliary circle.

And, similarly, if SQ, SR, perpendiculars to SB', SC', meet $C'A'$ and $A'B'$ in Q and R, these points correspond to BE and CF.

But AD, BE, CF are concurrent;

$$\therefore P, Q, R \text{ are collinear.}$$

Hence the reciprocal theorem,

If from any point S lines be drawn perpendicular respectively to SA', SB', SC', and meeting $B'C'$, $C'A'$, $A'B'$ in P, Q, and R, these points are collinear.

As a second illustration take the theorem,

If A, B be two fixed points, and AC, BC at right angles to each other, the locus of C is a circle.

Taking O, the middle point of AB, as the centre of the auxiliary circle, the reciprocals of A and B are two parallel straight lines, PE, QF, perpendicular to AB; the reciprocals of AC, BC are points P, Q on these lines such that POQ is a right angle, and PQ is the reciprocal of C.

Hence, the locus of C being a circle, it follows that PQ always touches a circle.

The reciprocal theorem therefore is,

If a straight line PQ, bounded by two parallel straight lines, subtend a right angle at a point O, halfway between the lines, the line PQ always touches a circle, having O for its centre.

223. PROP. II. *The reciprocal polar of a circle with regard to another circle, called the auxiliary circle, is a conic, a focus of which is the centre of the auxiliary circle, and the corresponding directrix the polar of the centre of the reciprocated circle.*

Let S be the centre of the auxiliary circle, and KX the polar of C, the centre of the reciprocated circle.

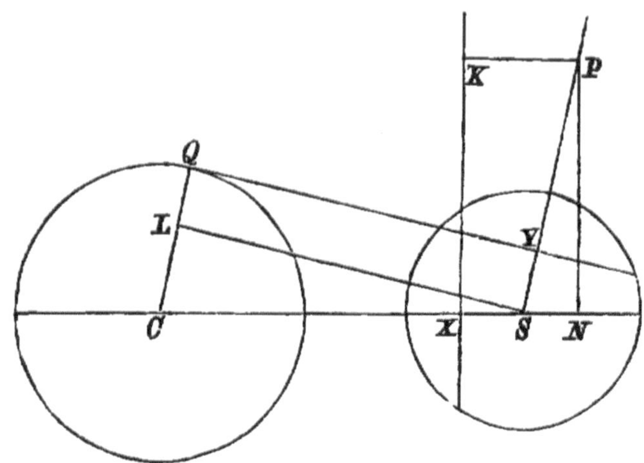

Then, if P be the pole of a tangent QY to the circle C, SP meeting this tangent in Y,

$$SP \,.\, SY = SX \,.\, SC.$$

Therefore, drawing SL parallel to QY,

$$SP : SC :: SX : QL.$$

But, by similar triangles,

$$SP : SC :: SN : CL;$$

$$\therefore SP : SC :: NX : CQ,$$

or

$$SP : PK :: SC : CQ.$$

Hence the locus of P is a conic, focus S, directrix KX and having for its eccentricity the ratio of SC to CQ.

The reciprocal polar of a circle is therefore an ellipse, parabola, or hyperbola, as the point S is within, upon, or without the circumference of the circle.

224. PROP. III. *To find the latus rectum and axes of the reciprocal conic.*

The ends of the latus rectum are the poles of the tangents parallel to SC.

Hence, if SR be the semi-latus rectum,

$$SR \,.\, CQ = SE^2,$$

SE being the radius of the auxiliary circle.

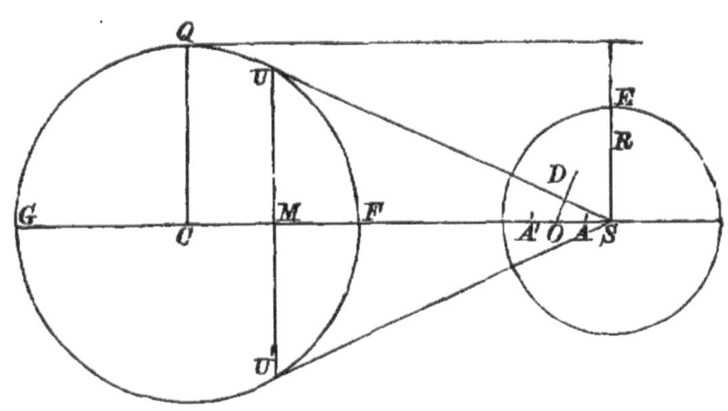

The ends of the transverse axis A, A' are the poles of the tangents at F and G;

$$\therefore SA\,.\,SG = SE^2$$

and $$SA'\,.\,SF = SE^2.$$

Let SU, SU' be the tangents from S, then

$$SG\,.\,SF = SU^2,$$

$$\left.\begin{aligned} \therefore SA' : SG &:: SE^2 : SU^2 \\ \text{and} \quad SA : SF &:: SE^2 : SU^2 \end{aligned}\right\} \quad (\alpha).$$

Hence $$AA' : FG :: SE^2 : SU^2,$$

or, if O be the centre of the reciprocal,

$$AO : CQ :: SE^2 : SU^2.$$

Again, if BOB' be the conjugate axis,

$$BO^2 = SR\,.\,AO;$$

therefore, since $$SE^2 = SR\,.\,CQ,$$

$$\begin{aligned} BO^2 : SE^2 &:: AO : CQ \\ &:: SE^2 : SU^2 \end{aligned}$$

and $$BO\,.\,SU = SE^2.$$

The centre O, it may be remarked, is the pole of UU'.

For, from the relations (α),

$$SE^2 : SU^2 :: SA + SA' : SF + SG$$
$$:: SO : SC$$
$$:: SO \,.\, SM : SC \,.\, SM;$$
$$\therefore SO \,.\, SM = SE^2.$$

225. In the figures drawn in the two preceding articles, the reciprocal conic is an hyperbola; the asymptotes are therefore the lines through O perpendicular to SU and SU', the poles of these lines being at an infinite distance.

The semi-conjugate axis is equal to the perpendicular from the focus on the asymptote (Art. 103), *i.e.* if OD be the asymptote, SD is equal to the semi-conjugate axis.

Further, since OD is perpendicular to SU, and O is the pole of UU', it follows that D is the pole of CU, and that

$$SD \,.\, SU = SE^2,$$

as we have already shewn.

Again, D, being the intersection of the polars of C and U, is the intersection of SU and the directrix.

226. If the point S be within the circle, so that the reciprocal is an ellipse, the axes are given by similar relations.

Through S draw SV perpendicular to FG, and let UMU' be the polar of S with regard to the circle.

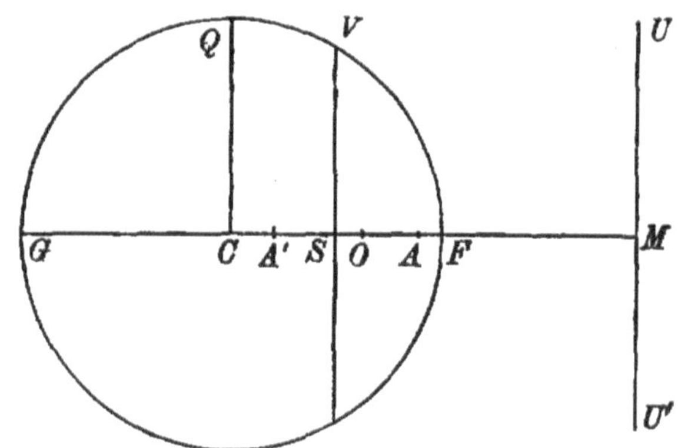

Then $SM . SC = SC . CM - SC^2 = CF^2 - SC^2 = SV^2$; also, SE being the radius of the auxiliary circle,

$$SA . SF = SE^2 = SA' . SG,$$

and
$$SF . SG = SV^2;$$

$$\left.\begin{array}{l} \therefore SA : SG :: SE^2 : SV^2 \\ SA' : SF :: SE^2 : SV^2 \end{array}\right\}.$$

Hence
$$SO : SC :: SE^2 : SV^2,$$

and
$$SO . SM : SC . SM :: SE^2 : SV^2;$$

$$\therefore SO . SM = SE^2,$$

so that O is the pole of UU'.

Again
$$SA + SA' : SF + SG :: SE^2 : SV^2,$$

$$\therefore AO : CQ :: SE^2 : SV^2.$$

If RSR' is the latus rectum,

$$SR . CQ = SE^2,$$

and if BOB' is the minor axis

$$SR . AO = BO^2;$$

$$\therefore BO^2 : SE^2 :: SE^2 : SV^2,$$

and
$$BO . SV = SE^2.$$

227. The important Theorem we have just considered enables us to deduce from any property of a circle a corresponding property of a conic, and we are thus furnished with a method, which may serve to give easy proofs of known properties, or to reveal new properties of conics.

In the process of reciprocation we observe that points become lines and lines points; that a tangent to a curve reciprocates into a point on the reciprocal, that a curve inscribed in a triangle becomes a curve circumscribing a triangle, and that when the auxiliary curve is a circle, the reciprocal of a circle is a conic, the latus rectum of which varies inversely as the radius of the circle.

Also, conversely, the reciprocal of a conic with regard to a circle having its centre at a focus of the conic is a circle the centre of which is the reciprocal of the directrix of the conic.

For an ellipse the centre of reciprocation is within the circle, for a parabola it is upon the circle, and for an hyperbola it is outside the circle.

228. We give some transformations of theorems as illustrations of the preceding articles.

THEOREM.	RECIPROCAL.
The line joining the points of contact of parallel tangents of a circle passes through the centre.	The tangents at the ends of a focal chord intersect in the directrix.
The angles in the same segment of a circle are equal.	If a moveable tangent of a conic meet two fixed tangents, the intercepted portion subtends a constant angle at the focus.
Two of the common tangents of two equal circles are parallel.	If two conics have the same focus, and equal latera recta, the straight line joining two of their common points passes through the focus.
The tangent at any point of a circle is perpendicular to the diameter through the point.	The portion of the tangent to a conic between the point of contact and the directrix subtends a right angle at the focus.
A chord of a circle is equally inclined to the tangents at its ends.	The tangents drawn from any point to a conic subtend equal angles at a focus.
If a chord of a circle subtend a constant angle at a fixed point on the curve, the chord always touches a circle.	If two tangents of a conic move so that the intercepted portion of a fixed tangent subtends a constant angle at the focus, the locus of the intersection of the moving tangents is a conic having the same focus and directrix.
If a chord of a circle pass through a fixed point, the rectangle contained by the segments is constant.	The rectangle contained by the perpendiculars from the focus on two parallel tangents is constant.
If two chords be drawn from a fixed point on a circle at right angles to each other, the line joining their ends passes through the centre.	If two tangents of a conic move so that the intercepted portion of a fixed tangent subtends a right angle at the focus, the two moveable tangents meet in the directrix.

THEOREM.	RECIPROCAL.
If a circle be inscribed in a triangle, the lines joining the vertices with the points of contact meet in a point.	If a triangle be inscribed in a conic the tangents at the vertices meet the opposite sides in three points lying in a straight line.
The sum of the reciprocals of the radii of the escribed circles of a triangle is equal to the reciprocal of the radius of the inscribed circle.	With a given point as focus, four conics can be drawn circumscribing a triangle, and the latus rectum of one is equal to the sum of the latera recta of the other three.
The common chord of two intersecting circles is perpendicular to the line joining their centres.	If two parabolas have a common focus, the line joining it to the intersection of the directrices is perpendicular to the common tangent.
If circles pass through two fixed points, the locus of their centres is a straight line.	If conics have a fixed focus and a pair of fixed tangents in common, the corresponding directrices all pass through a fixed point.
Two tangents to a conic at right angles to each other intersect on a fixed circle.	Chords of a circle which subtend a right angle at a fixed point all touch a conic of which that point is a focus.

229. PROP. IV. *A system of coaxal circles can be reciprocated into a system of confocal conics.*

Let X be the point at which the radical axis crosses the line of centres, and let E and S be the limiting points of the system.

Then XE is equal to the length of the tangent XD to any one of the circles, and, therefore, if A is the centre of this circle, AD is the tangent at D to the circle whose centre is X and radius XE.

Hence it follows that $AE \,.\, AS = AD^2$, shewing that UU', the polar of S with regard to the circle A, passes through E.

Reciprocating with regard to S, the centre of the reciprocal curve is the pole of UU', and is consequently fixed; and the conics are therefore confocal.

Hence, if we reciprocate with regard to either limiting point, we obtain confocal conics.

In the particular case in which the circles all touch the radical axis, we obtain confocal and co-axial parabolas.

230. Prop. V. *The reciprocal polar of a conic with regard to a circle, or with regard to any conic, is a conic.*

Taking any two tangents of the conic, their reciprocal polars are points on the reciprocal curve, and the reciprocal polar of their point of intersection is the chord joining the points.

Since only two tangents can be drawn from a point to a conic, it follows that the reciprocal curve is always intersected by a straight line in two points only.

It follows therefore that the reciprocal curve is a conic.

In reciprocating a conic with regard to a circle, the reciprocal polar is an ellipse, parabola, or hyperbola, according as the centre S of the circle is inside, upon, or outside the conic.

In the second case the axis of the parabola is parallel to the normal at the point S, and in the third case the asymptotes are perpendicular to the tangents which can be drawn from the point S to the conic.

When the auxiliary curve is a conic, centre S, the first of the preceding statements holds good.

When the point S is on the conic, the axis of the parabola is parallel to the diameter of the auxiliary conic, which is conjugate to the tangent at S.

When the point S is outside the conic, the asymptotes of the hyperbola are parallel to those diameters of the auxiliary conic which are conjugate to the straight lines through S touching the conic to be reciprocated.

The following cases will serve to illustrate the theorem of this article.

231. *The reciprocal polar of a parabola with regard to a point on the directrix is a rectangular hyperbola.*

For the two tangents from the point are at right angles to each other, and therefore the asymptotes are at right angles to each other.

232. *The reciprocal polar of an ellipse or hyperbola, with regard to its centre, is a similar curve turned through a right angle about the centre.*

If CY is the perpendicular on the tangent at P, and Q the reciprocal of the tangent, $CQ \,.\, CY$ is constant.

But $CY \,.\, CD$ is constant;

$$\therefore CQ \text{ varies as } CD,$$

and the reciprocal curve is the same as the original curve, or similar to it.

233. *The chords of a conic which subtend a right angle at a fixed point P of a conic all pass through a fixed point in the normal at P.*

Reciprocating with regard to P, the reciprocal curve is a parabola, the axis of which is parallel to the normal to the conic, and the reciprocal of the chord is the point of intersection of tangents at right angles to each other.

The locus of this point is the directrix of the parabola, and, being at right angles to the normal, it follows, on reciprocating backwards, that the chord passes through a fixed point E in the normal.

To find the position of the point E,
let C be the centre of the conic, CA, CB its semi-axes, and PNP' the double ordinate, and let the normal meet the axes in G and g.

Since CA and CB bisect the angle PCP' and its supplement,

$$C(BPAP') \text{ is an harmonic pencil;}$$

$\therefore PGEg$ is an harmonic range, so that PE is the harmonic mean between PG and Pg.

In the case of an hyperbola $EGPg$ is an harmonic range.

In the case of a parabola, E is the point of intersection of the normal with the diameter through P'.

234. *The chords of a conic which subtend a right angle at a fixed point O not on the conic all touch a conic of which that point is a focus.*

Reciprocating with regard to O, the reciprocal of the envelope of the chords is the director circle of a conic, and therefore, reciprocating backwards, it follows that the envelope of the chords is a conic of which O is a focus. This of course includes the preceding theorem as a particular case, the fact being that when O is on the conic the envelope of the chords is a conic, with a vertex and focus at E, flattened into a straight line.

235. *If the sides of a triangle are tangents to a parabola, the orthocentre of the triangle is on the directrix of the parabola.*

This theorem is at once obtained by reciprocating, with regard to the orthocentre of the triangle, the theorem, proved in Art. 143, that, if a rectangular hyperbola passes through the angular points of a triangle, it also passes through the orthocentre of the triangle.

EXAMPLES.

1. If any triangle be reciprocated with regard to its orthocentre, the reciprocal triangle will be similar and similarly situated to the original one and will have the same orthocentre.

2. If two conics have the same focus and directrix, and a focal chord be drawn, the four tangents at the points where it meets the conics intersect in the same point of the directrix.

3. An ellipse and a parabola have a common focus; prove that the ellipse either intersects the parabola in two points, and has two common tangents with it, or else does not cut it.

4. Prove that the reciprocal polar of the circumscribed circle of a triangle with regard to the inscribed circle is an ellipse, the major axis of which is equal in length to the radius of the inscribed circle.

5. Reciprocate with respect to any point S the theorem that, if two points on a circle be given, the pole of PQ with respect to that circle lies on the line bisecting PQ at right angles.

6. If two parabolas whose axes are at right angles have a common focus, prove that the part of the common tangent intercepted between the points of contact subtends a right angle at the focus.

7. The tangent at a moving point P of a conic intersects a fixed tangent in Q, and from S a straight line is drawn perpendicular to SQ and meeting in R the tangent at P; prove that the locus of R is a straight line.

8. Four parabolas having a common focus can be described touching respectively the sides of the triangles formed by four given points.

9. A triangle ABC circumscribes a parabola, focus S; through ABC lines are drawn respectively perpendicular to SA, SB, SC; shew that these lines are concurrent.

10. Prove that the distances, from the centre of a circle, of any two poles are to one another as their distances from the alternate polars.

11. Reciprocate the theorems,

(1) The opposite angles of any quadrilateral inscribed in a circle are equal to two right angles.

(2) If a line be drawn from the focus of an ellipse making a constant angle with the tangent, the locus of its intersection with the tangent is a circle.

12. The locus of the intersection of two tangents to a parabola which include a constant angle is an hyperbola, having the same focus and directrix.

13. Two ellipses having a common focus cannot intersect in more than two real points, but two hyperbolas, or an ellipse and hyperbola, may do so.

14. ABC is any triangle and P any point: four conic sections are described with a given focus touching the sides of the triangles ABC, PBC, PCA, PAB respectively; shew that they all have a common tangent.

15. TP, TQ are tangents to a parabola cutting the directrix respectively in X and Y; ESF is a straight line drawn through the focus S perpendicular to ST, cutting TP, TQ respectively in E, F; prove that the lines EY, XF are tangents to the parabola.

16. With the orthocentre of a triangle as focus, two conics are described touching a side of the triangle and having the other two sides as directrices respectively; shew that their minor axes are equal.

17. Two parabolas have a common focus S; parallel tangents are drawn to them at P and Q intersecting the common tangent in P' and Q'; prove that the angle PSQ is equal to the angle between the axes, and the angle $P'SQ'$ is supplementary.

18. ABC is a given triangle, S a given point; on BC, CA, AB respectively, points A', B', C' are taken, such that each of the angles ASA', BSB', CSC', is a right angle. Prove that A', B', C' lie in the same straight line, and that the latera recta of the four conics, which have S for a common focus, and respectively touch the three sides of the triangles ABC, $AB'C'$, $A'BC'$, $A'B'C$ are equal to one another.

19. A parabola and hyperbola have the same focus and directrix, and SPQ is a line drawn through the focus S to meet the parabola in P, and the nearer branch of the hyperbola in Q; prove that PQ varies as the rectangle contained by SP and SQ.

20. If two equal parabolas have the same focus, the tangents at points angularly equidistant from the vertices meet on the common tangent.

21. If an ellipse and a parabola have the same focus and directrix, and if tangents are drawn to the ellipse at the ends of its major axis, the diagonals of

the quadrilateral formed by the four points where these tangents cut the parabola intersect in the focus.

22. Find the reciprocals of the theorems of Arts. 215 and 217.

23. If a conic be reciprocated with regard to a point, shew that there are only two positions of the point, such that the conic may be similar and similarly situated to the reciprocal.

24. Conics are described having a common focus and equal latera recta. Also the corresponding directrices envelope a fixed confocal conic. Prove that these conics all touch two fixed conics, and that the reciprocals of the latera recta of these fixed conics are equal to the sum and difference of the latera recta of the variable conics and of the fixed confocal.

25. Given a point, a tangent, and a focus of a conic, prove that the envelope of the directrix is a conic passing through the given focus.

26. Two conics have a common focus: their corresponding directrices will intersect on their common chord, at a point whose focal distance is at right angles to that of the intersection of their common tangents.

If the conics are parabolas, the inclination of their axes will be the angle subtended by the common tangent at the common focus.

27. If the intercept on a given straight line between two variable tangents to a conic subtends a right angle at the focus of the conic, the tangents intersect on a conic.

28. The tangent at P to an hyperbola meets the directrix in Q; another point R is taken on the directrix such that QR subtends at the focus an angle equal to that between the transverse axis and an asymptote; prove that RP envelopes a parabola.

29. S is the focus of a conic; P, Q two points on it such that the angle PSQ is constant; through S, SR, ST are drawn meeting the tangents at P, Q in R, T respectively, and so that the angles PSR, QST are constant; shew that RT always touches a conic having the same focus and directrix as the original conic.

30. OA, OB are common tangents to two conics having a common focus S, CA, CB are tangents at one of their points of intersection, BD, AE tangents intersecting CA, CB, in D, E. Prove that SDE is a straight line.

31. An hyperbola, of which S is one focus, touches the sides of a triangle ABC; the lines SA, SB, SC are drawn, and also lines SD, SE, SF respectively perpendicular to the former three lines, and meeting any tangent to the curve in D, E, F; shew that the lines AD, BE, CF are concurrent.

32. If a conic inscribed in a triangle has one focus at the centre of the circumscribed circle of the triangle, its transverse axis is equal to the radius of that circle.

33. If any two diameters of an ellipse at right angles to each other meet the tangent at a fixed point P in Q and R, the other two tangents through Q and R intersect on a fixed straight line which passes through a point T on the tangent at P, such that PCT is a right angle.

CHAPTER XIII.

The Construction of a Conic from Given Conditions.

236. It will be found that, in general, five conditions are sufficient to determine a conic, but it sometimes happens that two or more conics can be constructed which will satisfy the given conditions. We may have, as given conditions, points and tangents of the curve, the directions of axes or conjugate diameters, the position of the centre, or any characteristic or especial property of the curve.

Prop. I. *To construct a parabola, passing through three given points, and having the direction of its axis given.*

In this case the fact that the conic is a parabola is one of the conditions.

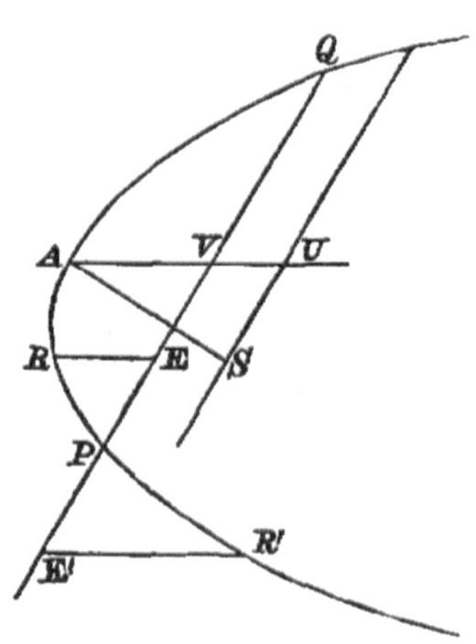

Let P, Q, R be the given points, and let RE parallel to the given direction meet PQ in E.

If E be the middle point of PQ, R is the vertex of the diameter RE; but, if not, bisecting PQ in V, draw the diameter through V and take A such that

$$AV : RE :: QV^2 : QE \,.\, EP.$$

Then A is the vertex of the diameter AV.

If the point E do not fall between P and Q, A must be taken on the side of PQ which is opposite to R.

The focus may then be found by taking AU such that

$$QV^2 = 4AV \,.\, AU,$$

and by then drawing US parallel to QV and taking AS equal to AU.

237. PROP. II. *To describe a parabola through four given points.*

First, let $ABCD$ be four points in a given parabola, and let the diameter CF meet AD in F.

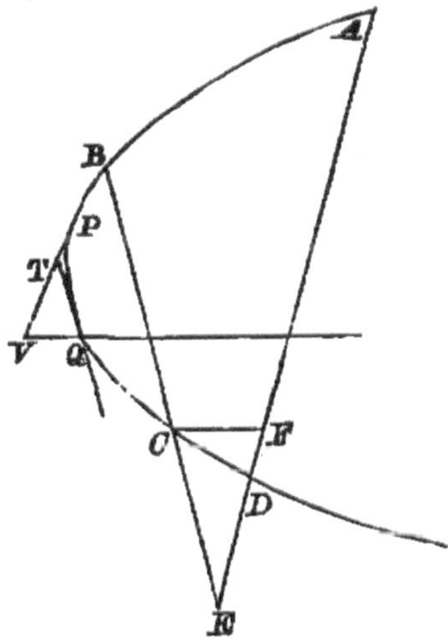

Draw the tangents PT, QT parallel to AD, BC, and the diameter QV meeting PT in V.

Then

$$\begin{aligned} ED \,.\, EA : EC \,.\, EB &:: TP^2 : TQ^2 \\ &:: TV^2 : TQ^2 \\ &:: EF^2 : EC^2. \end{aligned}$$

Hence the construction; in EA take EF such that

$$EF^2 : EC^2 :: ED \,.\, EA : EC \,.\, EB,$$

then CF is the direction of the axis, and the problem is reduced to the preceding.

If the point F be taken in AE produced, another parabola can be drawn, so that, in general, two parabolas can be drawn through four points.

238. This problem may be treated differently by help of the theorem of Art. 52, viz.;

If from a point O, outside a parabola, a tangent OM, and a chord OAB be drawn, and if the diameter ME meet the chord in E,

$$OE^2 = OA \,.\, OB.$$

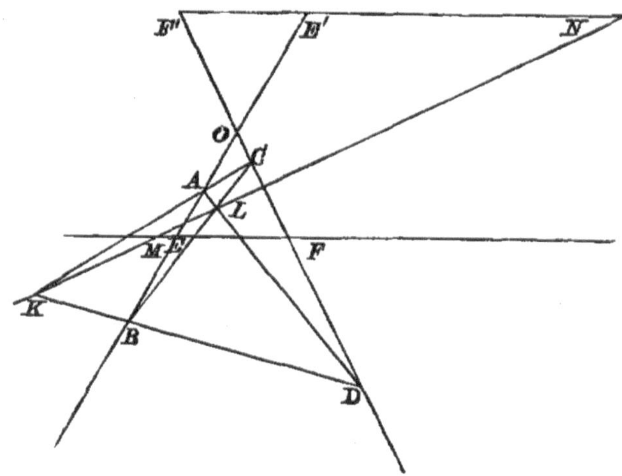

Let A, B, C, D be the given points, and let E, E', F, F', be so taken that

$$OE^2 = OE'^2 = OA \,.\, OB,$$

and

$$OF^2 = OF'^2 = OC \,.\, OD.$$

Then EF and $E'F'$ are diameters, and KL, the polar of O, will meet EF and $E'F'$ in M, N, the points of contact of tangents from O.

The second parabola is obtained by taking for diameters EF' and $E'F$.

239. PROP. III. *Any conic passing through four points has a pair of conjugate diameters parallel to the axes of the two parabolas which can be drawn through the four points.*

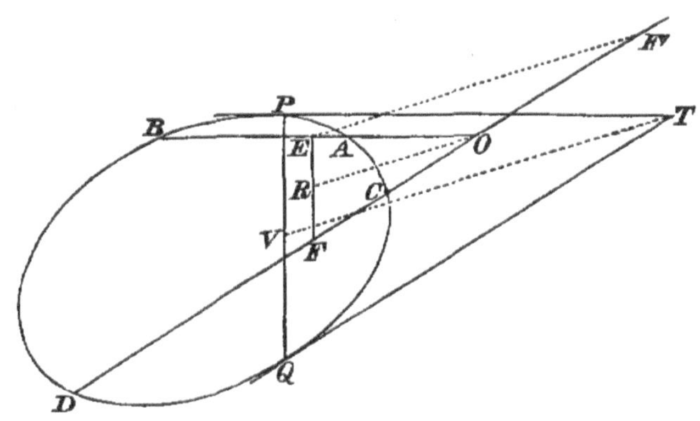

Let TP, TQ be the tangents parallel to OAB and OCD, and such that the angle PTQ is equal to AOC.

Then, if $OE^2 = OA . OB$, and $OF^2 = OC . OD$,

$$OE^2 : OF^2 :: OA . OB : OC . OD$$
$$:: TP^2 : TQ^2;$$

$\therefore EF$ is parallel to PQ.

Hence, if R and V be the middle points of EF and PQ, OR is parallel to TV;

But, taking OF' equal to OF, OR is parallel to EF',

$\therefore TV$ and PQ are parallel to EF' and EF;

i.e. the conjugate diameters parallel to TV and PQ are parallel to the axes of the two parabolas.

240. PROP. IV. *Having given a pair of conjugate diameters, PCP', DCD', it is required to construct the ellipse.*

In CP take E such that $PE . PC = CD^2$, draw PF perpendicular to CD, and take FC' equal to FC.

About CEC' describe a circle, cutting PF in G and G'; then

$$PG . PG' = PE . PC = CD^2,$$

and GCG' is a right angle; therefore CG and CG' are the directions of the axes and their lengths are given by the relations,

$$PG \,.\, PF = BC^2,$$
$$PG' \,.\, PF = AC^2.$$

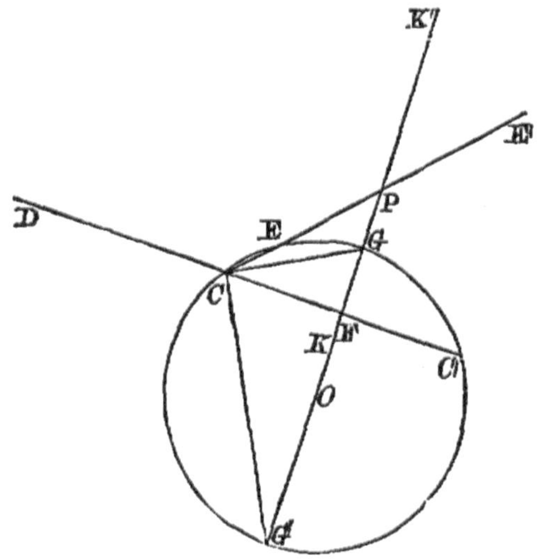

We may observe that, O being the centre of the circle,

$$\begin{aligned} AC^2 + BC^2 &= PF \,.\, PG + PF \,.\, PG' \\ &= 2 \,.\, PF \,.\, PO \\ &= 2 \,.\, PC \,.\, PN, \end{aligned}$$

if N be the middle point of CE,

$$\begin{aligned} &= PC^2 + PC \,.\, PE \\ &= CP^2 + CD^2. \end{aligned}$$

If PE' be taken equal to PE in CP produced, and the same construction be made, we shall obtain the axes of an hyperbola having CP, CD for a pair of conjugate semi-diameters.

241. This problem may be treated also as follows.

In PF, the perpendicular on CD, take

$$PK = PK' = CD;$$

then
$$PK^2 = PG \,.\, PG',$$

and therefore $K'GKG'$ is an harmonic range; and GCG' being a right angle, it follows (Art. 199), that CG and CG' are the bisectors of the angles between CK and CK'.

Hence, knowing CP and CD, G and G' are determined.

242. PROP. V. *Having given the focus and three points of a conic, to find the directrix.*

Let A, B, C, S be the three points and the focus.

Produce BA to D so that

$$BD : AD :: SB : SA,$$

and CB to E, so that

$$BE : CE :: SB : SC;$$

then DE is the directrix.

The lines BA, BC may be also divided internally in the same ratio, so that four solutions are generally possible.

Conversely, if three points A, B, C and the directrix are given, let BA, BC meet the directrix in D and E; then S lies on a circle, the locus of a point, the distances of which from A and B are in the ratio of AD to DB.

S lies also on a circle, similarly constructed with regard to BCE; the intersection of these circles gives two points, either of which may be the focus.

243. PROP. VI. *Having given the centre, the directions of a pair of conjugate diameters, and two points of an ellipse, to describe the ellipse.*

If C be the centre, CA, CB the given directions, and P, Q the points, draw QMQ', PLP' parallel to CB and CA, and make $Q'M = QM$ and $P'L = PL$.

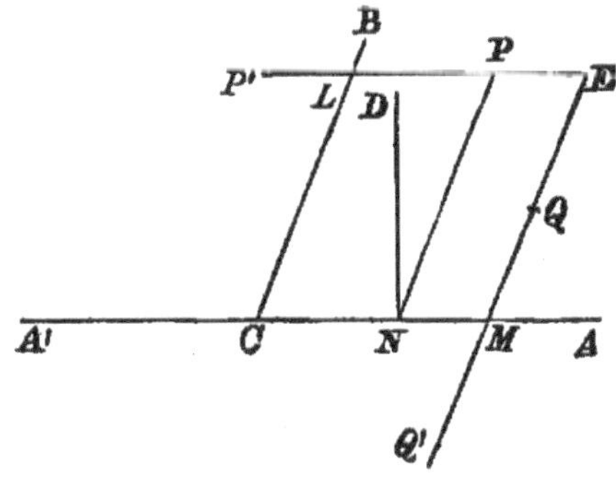

Then the ellipse will evidently pass through P' and Q', and if CA, CB be the conjugate radii, their ratio is given by the relation

$$CA^2 : CB^2 :: EP \,.\, EP' : EQ \,.\, EQ',$$

E being the point of intersection of $P'P$ and $Q'Q$.

Set up a straight line ND perpendicular to CA and such that

$$ND^2 : NP^2 :: EP \,.\, EP' : EQ \,.\, EQ',$$

and describe a circle, radius CD and centre C, cutting CA in A, and take

$$CB : CA :: NP : ND.$$

Then $$AN \,.\, NA' = ND^2,$$

and $$PN^2 : AN \,.\, NA' :: CB^2 : CA^2.$$

Hence CA, CB are determined, and the ellipse passes through P and Q.

244. Prop. VII. *To describe a conic passing through a given point and touching two given straight lines in given points.*

Let OA, OB be the given tangents, A and B the points of contact, N the middle point of AB.

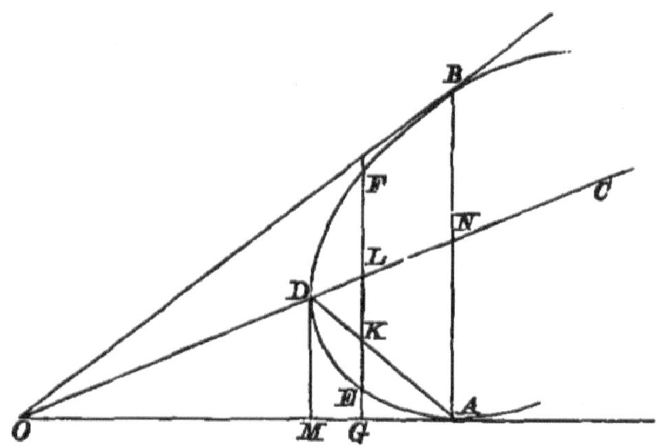

1st. Let the given point D be in ON; then, if $ND = OD$, the curve is a parabola.

But if $ND < OD$, the curve is an ellipse, and, taking C such that $OC \,.\, CN = CD^2$, the point C is the centre.

If $ND > OD$, the curve is an hyperbola, and its centre is found in the same manner.

2nd. If the given point be E, not in ON, draw GEF parallel to AB, and make FL equal to EL.

Take K such that

$$GK^2 = GE \,.\, GF;$$

then AK produced will meet ON in D, and the problem is reduced to the first case.

To justify this construction, observe that, if DM be the tangent at D,

$$GE \,.\, GF : GA^2 :: DM^2 : MA^2$$
$$:: GK^2 : GA^2,$$

so that $$GE \,.\, GF = GK^2.$$

245. Prop. VIII. *To draw a conic through five given points.*

Let A, B, C, D, E be the five points, and F the intersection of DE, AB.

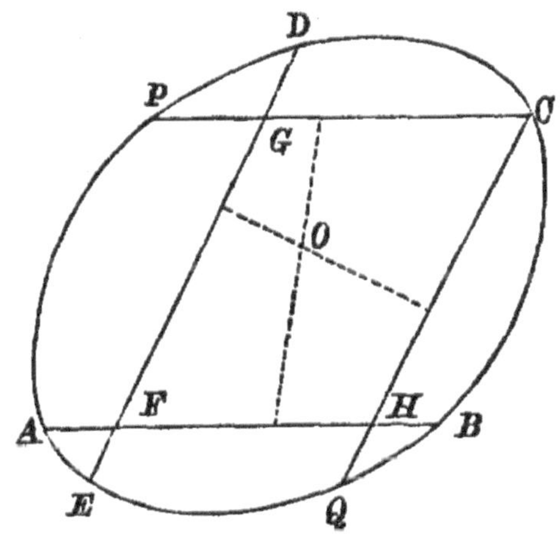

Draw CG, CH, parallel respectively to AB and ED, and meeting ED, AB in G and H.

If F and G fall between D and E, and F and H between A and B, take GP in CG produced and HQ in CH produced, such that

$$CG \,.\, GP : DG \,.\, GE :: AF \,.\, FB : DF \,.\, FE,$$

and $$CH \,.\, HQ : AH \,.\, HB :: DF \,.\, FE : AF \,.\, FB;$$

Then (Arts. 92 and 134) P and Q are points in the conic.

Also PC, AB being parallel chords, the line joining their middle points is a diameter, and another diameter is obtained from CQ and DE.

If these diameters are parallel, the conic is a parabola, and we fall upon the case of Prop. II.; but if they intersect in a point O, this point is the centre of the conic, and, having the centre, the direction of a diameter, and two ordinates of that diameter, we fall upon the case of Prop. VI.

The figure is drawn for the case in which the pentagon $AEBCD$ is not re-entering, in which case the conic may be an ellipse, a parabola, or an hyperbola.

If any one point fall within the quadrilateral formed by the other four, the curve is an hyperbola.

In all cases the points P, Q must be taken in accordance with the following rule.

The points C, P, or C, Q must be on the same or different sides of the points G, or H, according as the points D, E, or B, A are on the same or different sides of the points G or H.

Thus, if the point E be between D and F, and if G be between D and E, and H between A and B, the points P and C will be on the same side of G, and C, Q on the same side of H, but if H do not fall between A and B, C and Q will be on opposite sides of H.

Remembering that if a straight line meet only one branch of an hyperbola, any parallel line will meet only one branch, and that if it meet both branches, any parallel will meet both branches, the rule may be established by an examination of the different cases.

246. The above construction depends only on the elementary properties of Conics, which are given in Chapters I., II., III., and IV. For some further constructions we shall adopt another method depending on harmonic properties.

Prop. IX. *Having given two pairs of lines OA, OA', and OB, OB', to find a pair of lines OC, OC', which shall make with each of the given pairs an harmonic pencil.*

This is at once effected by help of Art. 203.

For, if any transversal cut the lines in the points c, a, b, c', b', a', the points c, c' are the foci of the involution, in which a, a' are conjugate, and also b, b', the centre of the involution being the middle point of cc'.

247. PROP. X. *If two points and two tangents of a conic be given, the chord of contact intersects the given chord in one of two fixed points**.

Let OP, OQ be the given tangents, A and B the given points, and C the intersection of AB and the chord of contact.

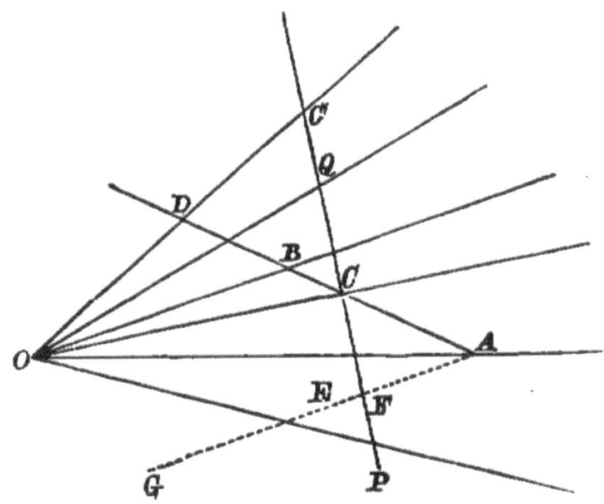

Let OC' be the polar of C, and let AB meet OC' in D.

Then C is on the polar of D, and therefore $DBCA$ is an harmonic range.

Also, C being on the polar of C', $C'QCP$ is an harmonic range.

Hence if two lines OC, OC' be found, which are harmonic with OA, OB, and also with OP, OQ, these lines intersect AB in two points C and D, through one of which the chord of contact must pass.

Or thus, if the tangents meet AB in a and b, find the foci C and D of the involution AB, ab; the chord of contact passes through one of these points.

248. PROP. XI. *Having given three points and two tangents, to find the chord of contact.*

In the preceding figure let OP, OQ be the tangents, and A, B, E the points.

Find OC, OC' harmonic with OA, OB, and OP, OQ; also find OF, OG harmonic with OA, OE and OP, OQ.

*I am indebted to Mr Worthington for much valuable assistance in this chapter, and especially for the constructions of Articles 247, 249, 250, and 253.

Then any one of the four lines joining C or D to F or G is a chord of contact, and the chord of contact and points of contact being known, the case reduces to that of Art. 244.

Hence four such conics can in general be described.

249. PROP. XII. *To describe a conic, passing through two given points, and touching three given straight lines.*

Let AB, the line joining the given points, meet the given tangents QR, RP, PQ, in N, M, L.

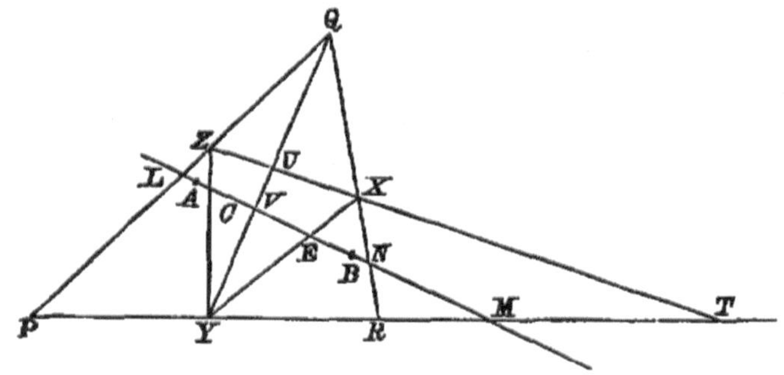

Find the foci C, D of the involution A, B and L, M;

Then YZ, the polar of P, passes through C or D, Art. 247.

Also find the foci, E, F, of the involution A, B, and M, N; then XY, the polar of R, passes through F or E.

Let ZX meet PR in T; then T is on the polar of Q, and QY is the polar of T.

Hence $TXUZ$ is harmonic;

therefore $MEVC$ is harmonic.

This determines V, and, joining QV, we obtain the point of contact Y.

Then, joining YC and YE, Z and X are obtained, and X, Y, Z being points of contact, we have five points, and can describe the conic by the construction of Art. 245, or by that of Art. 252.

Since either C or D may be taken with E or F, there are in general four solutions of the problem.

250. PROP. XIII. *To describe a conic, having given four points and one tangent.*

Let A, B, C, D be the given points, and complete the quadrilateral.

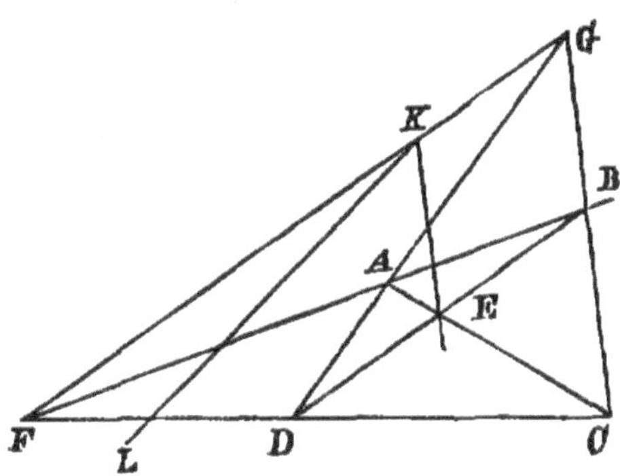

Then E is the pole of FG, and if the given tangent KL meet FG in K, E is on the polar of K; therefore the other tangent through K forms an harmonic pencil with KF, KL, KE.

Hence two tangents being known, and a point E in the chord of contact, if we find two points P, P' in A, B, such that KP, KP' are harmonic with KA, KB, and also with KL, KL', we shall have two chords of contact EP, EP', and therefore two points of contact for KL and also for KL'.

Hence two conics can be described.

We observe that if two conics pass through four points, their common tangents meet on one of the sides of the self-conjugate triangle EFG.

251. PROP. XIV. *Given four tangents and one point, to construct the conic.*

Let $ABCD$ be the given circumscribing quadrilateral, and E the given point. Completing the figure, draw LEF through E and F, and complete the harmonic range $LEFE'$; then, since F is the pole of HG (Art. 217), E' is a point in the conic.

Also, since K is the pole of FA (Art. 217), the chord of contact of the tangents AB, AD, passes through K.

Hence the construction is the same as that of Art. 250, and there are two solutions of the problem.

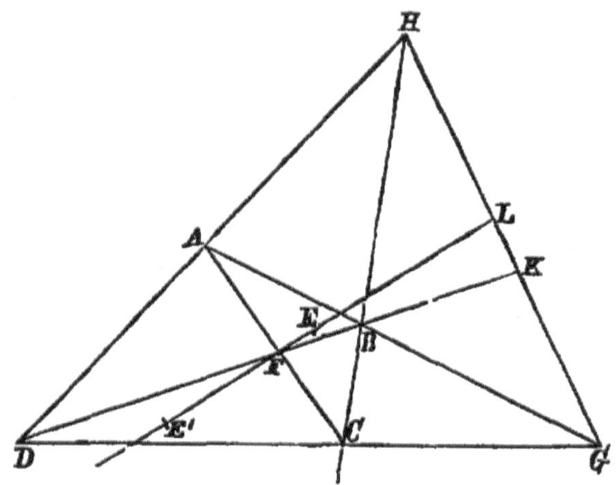

252. Prop. XV. *Given five points, to construct the conic.*

Let A, B, C, D, E be the five points, and complete the quadrilateral $ABCD$.

Then H is the pole of FG, and FG passes through the points of contact P, Q of the tangents from H.

Join HE, cutting FG in K, and complete the harmonic range $HEKE'$; then E' is a point in the conic.

Also AE, BE' will intersect FG in the same point F', and $E'A$, EB will also intersect FG in the same point G'.

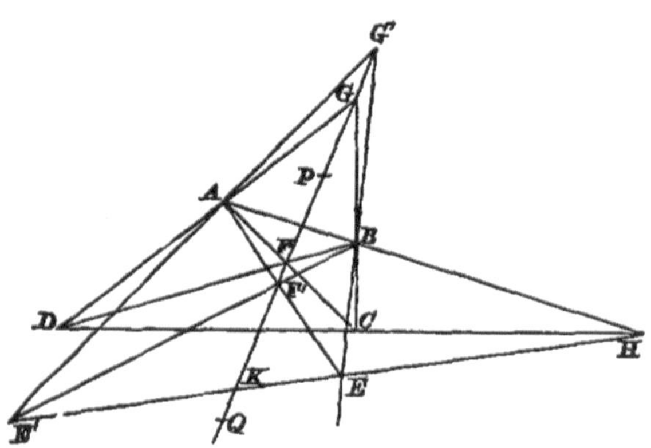

But $GPFQ$ and $G'PF'Q$ are both harmonic ranges, therefore P and Q are the foci of an involution of which F, G and F', G' are pairs of conjugate points.

Hence, finding these foci, P and Q, the tangents HP, HQ are known, and the case is reduced to that of Prop. VII.

Hence only one conic can be drawn through five points.

253. PROP. XVI. *Given five tangents, to find the points of contact.*

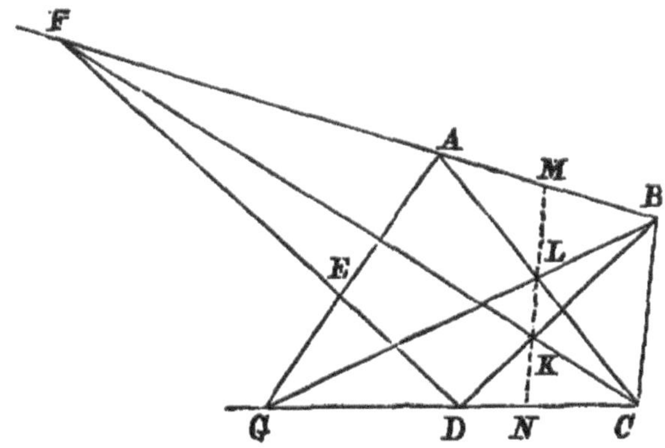

Let $ABCDE$ be the circumscribing pentagon. Considering the quadrilateral $FBCD$, join FC, BD, meeting in K.

Then (Art. 217) K is the pole of the line joining the intersections of FB, CD, and of FD, BC; that is, the chords of contact of BF, CD, and of BC, FD meet in K.

Similarly if BG, AC meet in L, the chords of contact of AB, CG, and of BC, AG meet in L.

Hence KL is the chord of contact of AB, CD, and therefore determines M, N the points of contact.

Hence it will be seen that only one conic can be drawn touching five lines.

CHAPTER XIV.

The Oblique Cylinder, the Oblique Cone, and the Conoids.

254. Def. If a straight line, which is not perpendicular to the plane of a given circle, move parallel to itself, and always pass through the circumference of the circle, the surface generated is called an oblique cylinder.

The line through the centre of the circular base, parallel to the generating lines, is the axis of the cylinder.

It is evident that any section by a plane parallel to the axis consists of two parallel lines, and that any section by a plane parallel to the base is a circle.

The plane through the axis perpendicular to the base is the principal section.

The section of the cylinder by a plane perpendicular to the principal section, and inclined to the axis at the same angle as the base, is called a subcontrary section.

255. Prop. I. *The subcontrary section of an oblique cylinder is a circle.*

The plane of the paper being the principal plane and APB the circular base, a subcontrary section is DPE, the angles BAE, DEA being equal.

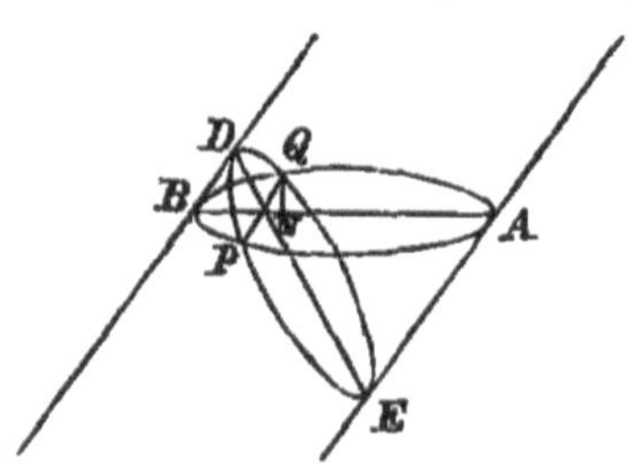

Let PQ be the line of intersection of the two sections; then

$$PN \,.\, NQ \text{ or } PN^2 = BN \,.\, NA.$$

But $\quad NB = ND$, and $NA = NE$;

$$\therefore PN \,.\, NQ = DN \,.\, NE,$$

and DPE is a circle.

256. Prop. II. *The section of an oblique cylinder by a plane which is not parallel to the base or to a subcontrary section is an ellipse.*

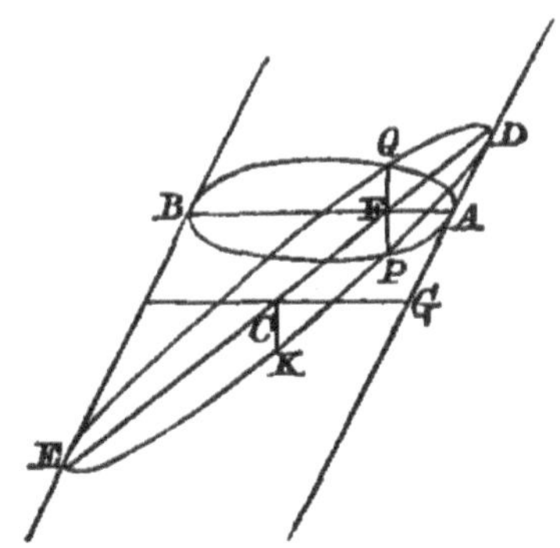

Let the plane of the section, DPE, meet any circular section in the line PQ, and let AB be that diameter of the circular section which is perpendicular to PQ, and bisect PQ in the point F.

Let the plane through the axis and the line AB cut the section DPE in the line DPE.

Then $$PF^2 = AF \,.\, FB.$$

But if DE be bisected in C, and GKC be the circular section through C parallel to APB,

$$AF : FD :: CG : CD,$$

and $$FB : FE :: CG : CD;$$

$$\therefore AF \,.\, FB : DF \,.\, FE :: CG2 : CD2;$$

hence, observing that $CG = CK$,

$$PF^2 : DF \,.\, FE :: CK^2 : CD^2.$$

But, if a series of parallel circular sections be drawn, PQ is always parallel to itself and bisected by DE;

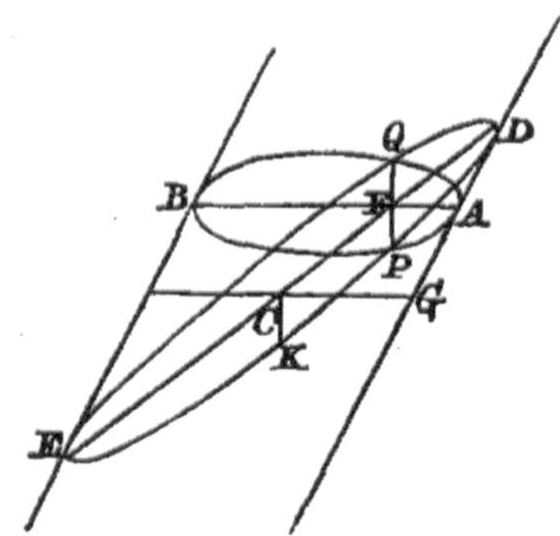

Therefore the curve DPE is an ellipse, of which CD, CK are conjugate semi-diameters.

257. DEF. If a straight line pass always through a fixed point and the circumference of a fixed circle, and if the fixed point be not in the straight line through the centre of the circle at right angles to its plane, the surface generated is called an oblique cone.

The plane containing the vertex and the centre of the base, and also perpendicular to the base, is called the principal section.

The section made by a plane not parallel to the base, but perpendicular to the principal section, and inclined to the generating lines in that section at the same angle as the base, is called a subcontrary section.

258. PROP. III. *The subcontrary section of an oblique cone is a circle.*

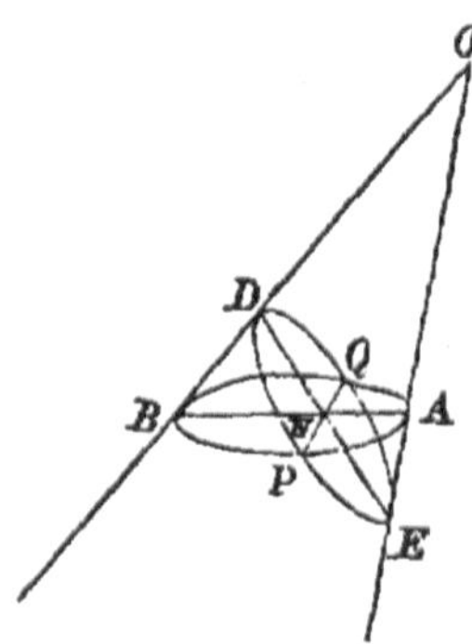

The plane of the paper being the principal section, let APB be parallel to the base and DPE a subcontrary section, so that the angle

$$ODE = OAB,$$

and

$$OED = OBA.$$

The angles DBA, DEA being equal to each other, a circle can be drawn through $BDAE$.

Hence, if PNQ be the line of intersection of the two planes APB and EPD,

$$\begin{aligned} DN \,.\, NE &= BN \,.\, NA, \\ &= PN \,.\, NQ; \end{aligned}$$

therefore DPE is a circle.

And all sections by planes parallel to DPE are circles.

Planes parallel to the base, or to a subcontrary section, are called also *Cyclic Planes*.

259. PROP. IV. *The section of a cone by a plane not parallel to a cyclic plane is an Ellipse, Parabola, or Hyperbola.*

(1) Let the section, DPE, meet all the generating lines on one side of the vertex.

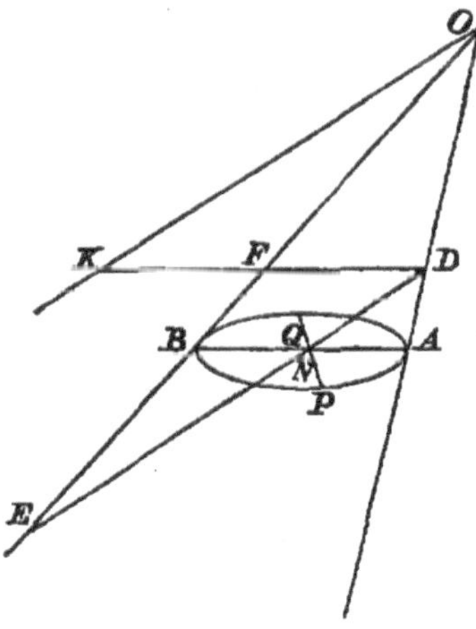

Let any circular section cut DPE in PQ, and take AB the diameter of the circle which bisects PQ.

The plane OAB will cut the plane of the section in a line DNE.

Draw OK parallel to DE and meeting in K the plane of the circular section through D parallel to APB, and join DK, meeting OE in F.

Then $$AN : ND :: KD : OK,$$

and $$BN : NE :: KF : OK;$$

therefore $$AN \,.\, NB : DN \,.\, NE :: KD \,.\, KF : OK^2,$$

or $$PN^2 : DN \,.\, NE :: KD \,.\, KF : OK^2.$$

But if a series of circular sections be drawn the lines PQ will always be parallel, and bisected by DE;

Therefore the curve DPE is an ellipse, having DE for a diameter, and the conjugate diameter parallel to PQ, and the squares on these diameters are in the ratio of $KD \,.\, KF$ to OK^2.

(2) Let the section be parallel to a tangent plane of the cone.

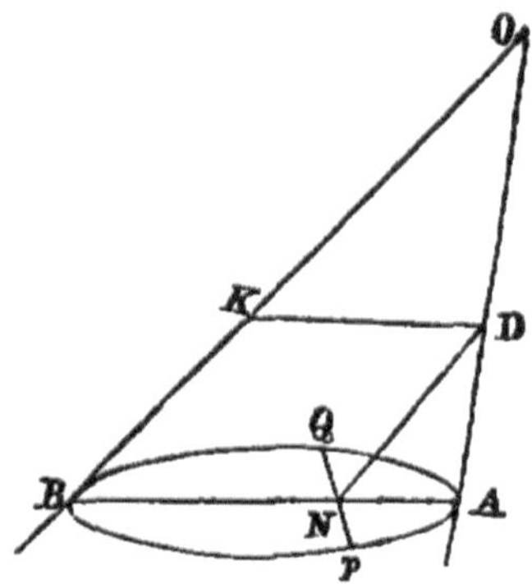

If OB be the generating line along which the tangent plane touches the cone, and BT the tangent line at B to a circular section through B, the line of intersection PQ will be parallel to BT, and therefore perpendicular to the diameter BA through B.

Let the plane BOA cut the plane of the section in DN.

Then, drawing DK parallel to AB,

$$BN = KD,$$

and $$AN : ND :: KD : OK;$$

therefore $$AN \,.\, NB : ND \,.\, KD :: KD : OK,$$

or $$PN^2 : ND \,.\, KD :: KD : OK,$$

and KD, OK being constant, the curve is a parabola having the tangent at D parallel to PQ.

If the plane of the section meet both branches of the cone, make the same construction as before, and we shall obtain, in the same manner as for the ellipse,

$$PN^2 : DN \,.\, NE :: DK \,.\, KF : OK^2,$$

OK being parallel to DE.

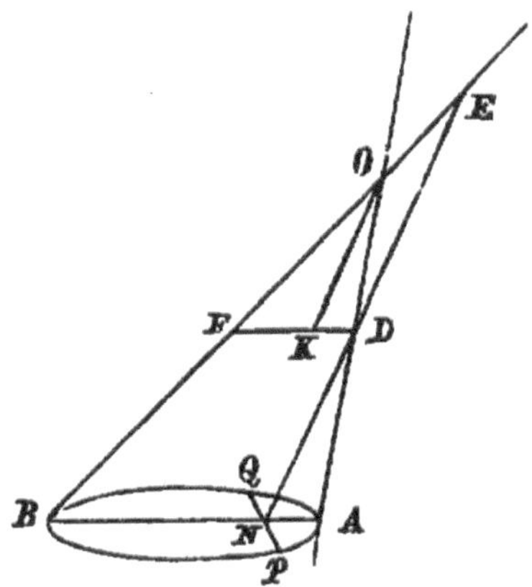

Therefore, since the point N is not between the points D and E, the curve DP is an hyperbola.

Conoids.

260. Def. *If a conic revolve about one of its principal axes, the surface generated is called a conoid.*

If the conic be a circle, the conoid is a sphere.

If the conic be an ellipse, the conoid is an oblate or a prolate spheroid according as the revolution takes place about the conjugate or the transverse axis.

If it be an hyperbola the surface is an hyperboloid of one or two sheets, according as the revolution takes place about the conjugate or transverse axis, and the surface generated by the asymptotes is called the asymptotic cone.

If the conic consist of two intersecting straight lines, the limiting form of an hyperbola, the revolution will be about one of the lines bisecting the angles between them, and the conoid will then be a right circular cone.

261. PROP. V. *A section of a paraboloid by a plane parallel to the axis is a parabola equal to the generating parabola, and any other section not perpendicular to the axis is an ellipse.*

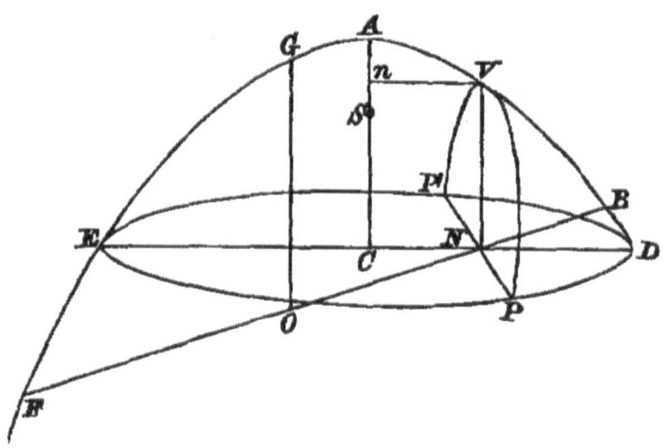

Let PVN be a section parallel to the axis, and take the plane of the paper perpendicular to the section and cutting it in VN.

Take any circular section DPE, cutting the section PVN in PNP'.

Then PN is perpendicular to DE,

and
$$\begin{aligned} PN^2 &= DN \,.\, NE \\ &= DC^2 - NC^2 \\ &= 4AS \,.\, AC - 4AS \,.\, An \\ &= 4AS \,.\, VN; \end{aligned}$$

therefore the curve VP is a parabola equal to EAD.

Again, let BPF be a section not parallel or perpendicular to the axis, but perpendicular to the plane of the paper;

Then, $BN \,.\, NF = 4SG \,.\, VN$, OG being the diameter bisecting BF (Art. 51);

therefore
$$PN^2 : BN \,.\, NF :: AS : SG,$$

and the curve BPN is an ellipse.

Moreover if the plane BF move parallel to itself, SG is unaltered, and the *sections by parallel planes are similar ellipses.*

In exactly the same manner, it may be shewn that the oblique sections of spheroids are ellipses, and those of hyperboloids either ellipses or hyperbolas.

262. PROP. VI. *The sections of an hyperboloid and its asymptotic cone by a plane are similar curves.*

Taking the case of an hyperboloid of two sheets, let DPF, $dP'f$, be the sections of the hyperboloid and cone, $P'PN$ the line in which their plane is cut by a circular section GPK or $gP'k$.

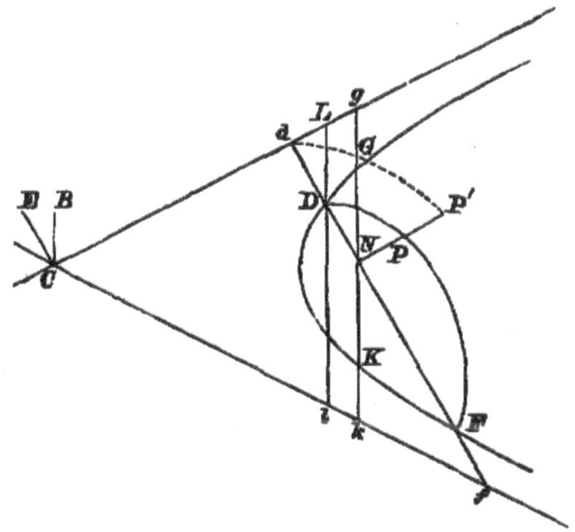

Through D draw LDl perpendicular to the axis; then, since

$$PN^2 = GN \,.\, NK, \text{ and } P'N^2 = gN \,.\, Nk,$$

$$P'N^2 : dN \,.\, Nf :: gN \,.\, Nk : dN \,.\, Nf,$$

$$:: LD \,.\, lD : Dd \,.\, Df,$$

$$:: BC^2 : CE^2$$

if CE be the semi-diameter parallel to DF;

and $$PN^2 : DN \,.\, NF :: GN \,.\, NK : DN \,.\, NF$$

$$:: BC^2 : CE^2 (\text{Art. } 134);$$

therefore the curves DPF, $dP'f$ have their axes in the same ratio, and are similar ellipses.

In the same manner the theorem can be established if the sections be hyperbolic, or if the hyperboloid be of one sheet.

263. PROP. VII. *If an hyperboloid of one sheet be cut by a tangent plane of the asymptotic cone, the section will consist of two parallel straight lines.*

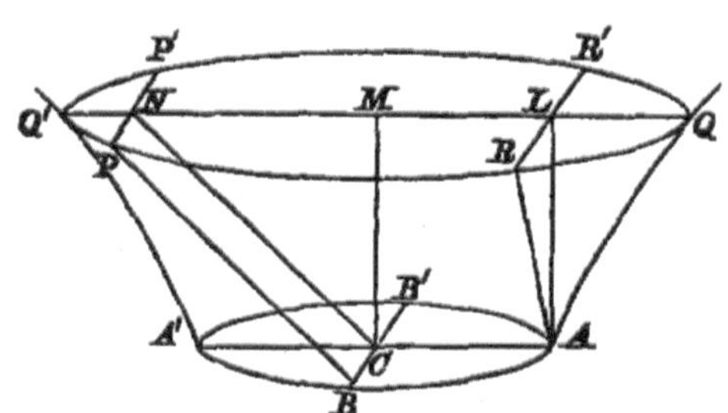

Let AQ, $A'Q'$ be a section through the axis, CN the generating line, in the plane CAQ, along which the tangent plane touches the cone; and PNP' the section with this tangent plane of a circular section QPQ'.

Then $$PN^2 = QN \,.\, NQ'$$
$$= AC^2 (\text{Art. } 106) = BC^2,$$
therefore, if BCB' be the diameter, perpendicular to the plane CAQ, of the principal circular section,
$$PN = BC \text{ and } P'N = B'C;$$
therefore PB and $P'B'$ are each parallel to CN; that is, the section consists of two parallel straight lines.

264. PROP. VIII. *The section of an hyperboloid of one sheet by a plane parallel to its axis, and touching the central circular section, consists of two straight lines.*

Let the plane pass through A, and be perpendicular to the radius CA of the central section (fig. Art. 263).

The plane will cut the circular section QPQ' in a line RLR', and
$$RL^2 = QL \,.\, LQ' = QM^2 - AC^2,$$
if M be the middle point of QQ'.

But $$QM^2 - AC^2 : CM^2 :: AC^2 : BC^2;$$
therefore $$RL : AL :: AC : BC;$$

hence it follows that AR is a fixed line; and similarly AR' is also a fixed line.

It will be seen that these lines are parallel to the section of the cone by the plane through the axis perpendicular to CA.

265. PROP. IX. *If a conoid be cut by a plane, and if spheres be inscribed in the conoid touching the plane, the points of contact of the spheres with the plane will be the foci of the section, and the lines of intersection of the planes of contact with the plane of section will be the directrices.*

In order to establish this statement, we shall first demonstrate the following theorem;

If a circle touch a conic in two points, the tangent from any point of the conic to the circle bears a constant ratio to its distance from the chord of contact.

Take the case of an ellipse, the chord of contact being perpendicular to the transverse axis.

If EME' be this chord, the normal EG is the radius of the circle, and if PT be a tangent from a point P of the ellipse,

$$PT^2 = PG^2 - GE^2$$
$$= PN^2 + NG^2 - EM^2 - MG^2.$$

But $$EM^2 - PN^2 : CN^2 - CM^2 :: BC^2 : AC^2,$$

and $$CN^2 - CM^2 = MN(CM + CN).$$

Let the normal at P meet the axis in G';

then $$NG' : CN :: BC^2 : AC^2,$$

and $$MG : CM :: BC^2 : AC^2;$$

therefore $$NG' + MG : CN + CM :: BC^2 : AC^2.$$

Hence $$EM^2 - PN^2 = MN(NG' + MG).$$

Also $$NG^2 - MG^2 = MN(NG + MG);$$

therefore $$PT^2 = MN(NG + MG) - MN(NG' + MG)$$
$$= MN \,.\, GG'.$$

But $$CG : CM :: SC^2 : AC^2,$$

and $$CG' : CN :: SC^2 : AC^2;$$

therefore $$GG' : MN :: SC^2 : AC^2.$$

Hence $$PT^2 : PL^2 :: SC^2 : AC^2,$$

PL being equal to MN.

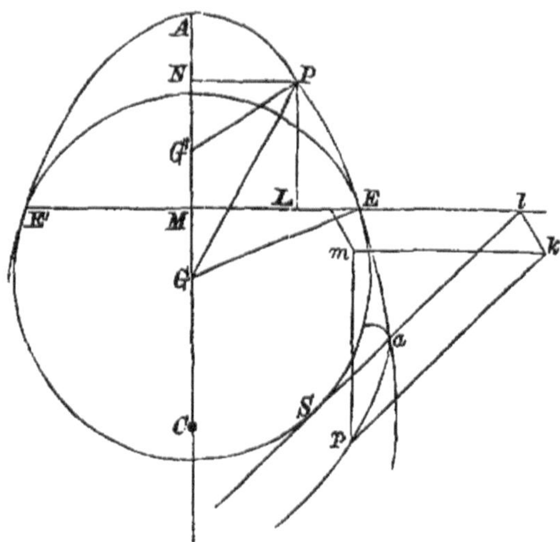

This being established let the figure revolve round the axis AC, and let a plane section ap of the conoid, perpendicular to the plane of the paper, touch the sphere at S and cut the plane of contact EE' in lk.

From a point p of the section let fall the perpendicular pm on the plane EE', draw mk perpendicular to lk, and join pk.

Then $pm : pk$ is a constant ratio.

Also taking the meridian section through p, pS is equal to the tangent from p to the circular section of the sphere, and is therefore in a constant ratio to pm;

Hence Sp is to pk in a constant ratio,
and therefore S is the focus and kl the directrix of the section ap.

266. If the curve be a parabola focus S', the proof is as follows:

$$\begin{aligned} PT^2 &= PG^2 - EG^2 \\ &= PN^2 + NG^2 - EM^2 - MG^2 \\ &= MN(NG + MG) - 4AS' \,.\, MN \\ &= MN(NG + MG) - 2MG \,.\, MN \\ &= MN^2. \end{aligned}$$

It will be found that the theorem is also true for an hyperboloid of two sheets, and for an hyperboloid of one sheet, but that in the latter case the constant ratio of PT to PL is not that of SC to AC.

267. The geometrical enunciation of the theorem also requires modification in several cases. To illustrate the difficulty, take the paraboloid, and observe that if the normal at E cuts the axis in G, and if O be the centre of curvature at A,

$$AG > AO,$$

and the radius of the circle is never less than AO.

This shews that a circle the radius of which is less than AO cannot be drawn so as to touch the conic in two points.

We may mention one exceptional case in which the theorem takes a simple form.

In general

$$EG^2 = EM^2 + MG^2 = 4AS'(AM + AS')$$
$$= 4AS' \,.\, S'G.$$

Taking the point g between S' and O, describe a circle centre g and such that the square on its radius $= 4AS' \,.\, S'g$.

Also take a point F in the axis produced such that

$$AF = Og;$$

it will then be found that the tangent from P to the circle will be equal to NF.

When g coincides with S', the circle becomes a point,

and $$AF = AS';$$

we thus fall back on the fundamental definition of a parabola.

It will be found that if the plane section of the conoid pass through S', the point S' is a focus of the section.

CHAPTER XV.

Conical Projection.

268. If from any fixed point straight lines are drawn to all the points of a figure, the section by any plane of the lines thus drawn is the conical projection of the figure upon that plane.

The fixed point is called the vertex of projection, and the plane is called the plane of projection.

Taking the eye as the vertex of projection, the conical projection of any figure upon a plane is a perspective drawing of that figure as seen by the eye.

A straight line is projected into a straight line, for the plane through the vertex and the straight line intersects the plane of projection in a straight line.

A tangent to a curve is projected into a tangent to the projection of the curve, for two consecutive points of a curve project into two consecutive points.

Hence it follows that *a pole and polar project into a pole and polar.*

Again, *the degree of a curve is unaltered by projection*, for any number of collinear points project into the same number of collinear points.

In particular, the projection of a conic on any plane is a conic.

269. *Any straight line in a figure can be projected to an infinite distance.*

This is effected by taking the plane of projection parallel to the plane through the vertex of projection and the straight line.

270. *A system of concurrent straight lines in a plane can be projected into a system of parallel straight lines, and a system of parallel straight lines can be projected into a system of concurrent straight lines.*

The first of these is effected by taking for plane of projection any plane parallel to the straight line joining the vertex of projection and the point of concurrence.

The second is effected by taking for plane of projection any plane not parallel to the direction of the parallel straight lines.

271. *Any angle in a plane can be projected, on any other plane, into any other angle.*

Let ACB be the angle to be projected, and let DEF be the plane upon which it is to be projected.

Take any plane parallel to DEF, intersecting in A and B the lines forming the angle ACB, and take any point O in the plane.

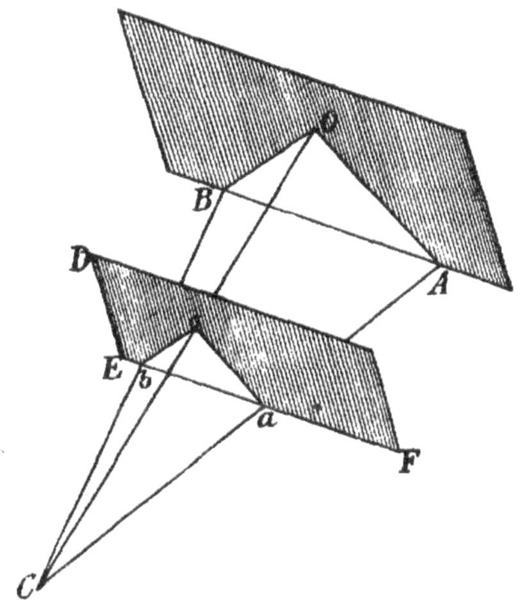

Then, if CA, CB, CO, meet the plane of projection in a, b, c, the angle acb is the projection of the angle ACB from the vertex O upon the plane DEF.

Now OA, OB are parallel to ca, cb; therefore the angle acb is equal to the angle AOB.

If then we describe on AB an arc of a circle containing an angle equal to any given angle, and take any point O on the arc as vertex of projection, the angle ACB will be projected into the given angle.

It will be seen that the arc of a circle may be described on the other side of the plane CAB, so that the locus of O on the plane OAB consists of two equal arcs on the same base.

If the plane of projection be assigned, it follows, since the plane OAB may be taken at any distance from C, that the locus of O consists of portions of two oblique cones having their common vertex at C.

If the plane of projection be not assigned, but if the line AB be assigned, the locus of O will be the surface generated by the revolution, about AB, of the arc of the circle.

If the angle ACB is to be projected into a right angle, the locus of O will be the sphere described upon AB as diameter.

If the assigned plane, DEF, be parallel to CA, the locus of O on the plane OBA will be the straight line BO making with BA the angle OBA equal to the supplement of the angle into which ACB is to be projected.

In the particular case in which this angle is a right angle the locus of O will be the straight line BO perpendicular to BA.

If it be required to project two given angles in a plane into two other given angles in any other plane, we can construct two arcs of circles in a plane parallel to this other plane, and, if these arcs intersect, the position of O is determined.

272. *To project a given quadrilateral into a square.*

Let $ABCD$ the quadrilateral, and let AC, BD intersect in E, AD, BC in F, and BA, CD in G.

Then if O is the vertex of projection, taken anywhere, the quadrilateral will be projected into a parallelogram on any plane parallel to OFG.

If O be taken on the sphere of which FG is diameter, the projection on any plane parallel to OFG will be a rectangle, for the angles subtended by FG at A, B, C, D project into right angles.

If AC and BD meet FG in L and M, and if O be taken on the circle which is the intersection of the spheres on FG and LM as diameters, the angle LEM will be projected into a right angle, so that the projection of $ABCD$ will be a rectangle, the diagonals of which are at right angles, and therefore will be a square.

273. *The projection of an harmonic range is an harmonic range.*

This is proved in Art. 198.

The projection of a circle is a conic.

This is proved in Art. 259.

As an illustration it is easily shown for a circle that, if a diameter pP passes through an external point T and intersects in V the polar of T, $pVPT$ is an harmonic range.

By projection we at once obtain the theorems of Art. 78 and of Art. 117.

274. *To project a conic into a circle, so that the projection of a given point inside the conic shall be the centre of the projection.*

Let E be the given point, AEB the chord bisected at E and PEp the diameter passing through E.

Then, if we project the polar of E to an infinite distance, and the angles AEP, APB into right angles, the projection of the conic will be a circle, the centre of which is the projection of the point E.

For the centre is the pole of a line at an infinite distance, and, the projection of AEP being a right angle, the projections of AB and Pp are the principal axes of the projection.

Also, the projection of APB being a right angle, it follows that the projection of the conic is a circle.

Another method will be to take points C, C', D, D' on the polar of E, such that CED, $C'ED'$ are self-conjugate triangles, and then to project CD to an infinite distance and the angles CED, $C'ED'$ into right angles.

The projection will be a conic, having the projection of E for its centre, and also having two pairs of conjugate diameters at right angles to each other; that is, it will be a circle.

In a subsequent article this question will be treated in a different manner.

If the point E is outside the conic, we can project the conic into a rectangular hyperbola, of which the projection of E is the centre.

For, if PQ is the chord of contact of tangents from E, all we have to do is to project PQ to an infinite distance, and PEQ into a right angle.

We can also project the conic into an hyperbola of any given eccentricity.

For, if the eccentricity is given, the angle between the asymptotes is given, and we can project PQ to an infinite distance and PEQ into the given angle.

275. *To project a conic on a given plane so that the projection of a point S inside the conic shall be a focus of the projection.*

Let the tangent at any point P and any straight line through S meet the polar of S in F and X.

Then, if we project the angles SXF, FSP into right angles, the projections of S and FX are the focus and directrix of the projection.

If at the same time we project to an infinite distance the polar of any point E on XS, the projection of E will be the centre of the projection of the conic.

276. *If two conics in different planes have two points in common, two cones of the second order can be drawn passing through them, or, in other words, each can be projected into the other.*

Let AB be the common chord, F and D its poles with regard to the conics.

Take any point E in AB, and let the plane FED meet the conics in the points P, p, Q, q, and let pq intersect DF in O.

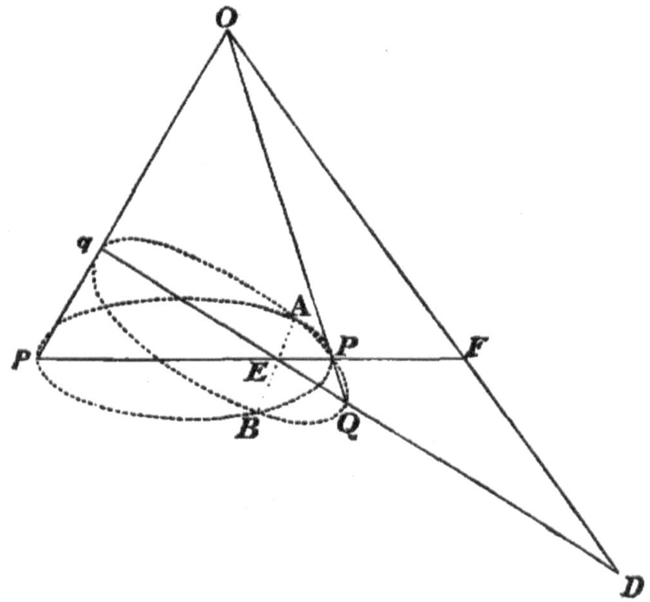

If from O the conic $BPAp$ be projected on to the plane of the other conic, the projection will be a conic touching the conic $BQAq$ at A and B, so that it will have four points in common with $BQAq$, and will also have the point q in common with $BQAq$.

Now it is proved in Art. 252, that only one conic can be drawn through five points.

Hence the projection, having five points in common with $BQAq$, coincides with it entirely.

It will be observed that OPQ is a straight line, Pp being projected into Qq.

The point O is therefore the vertex of a quadric cone which passes through the two conics.

The vertex of another such cone is obtained by producing qP or pQ to meet DF.*

277. *A conic can be projected into a circle so that the projection of any point inside the conic shall be the centre of the circle.*

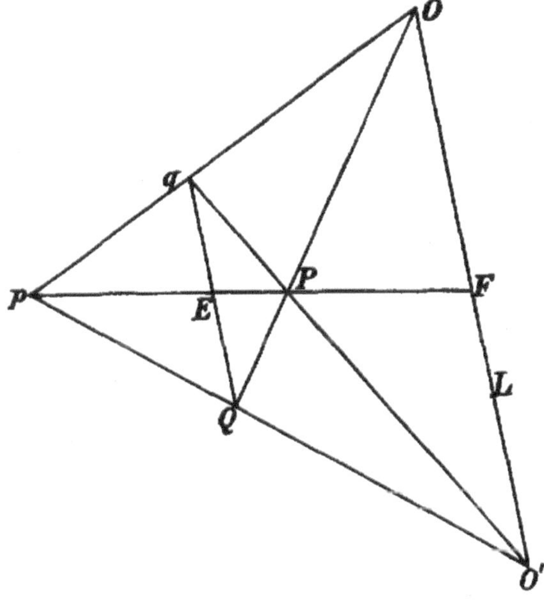

Let E be the point inside the conic and let AB be the chord of which E is the middle point.

*I am indebted to Mr H. F. Baker, Fellow and Lecturer of St John's College, for having called my attention to this theorem and to the mode of proof which is here given. The theorem is given in Poncelet's Treatise, and also in the article on Projections in the last edition of the *Encyclopaedia Britannica*.

Describe a circle on AB as diameter in any plane passing through AB.

Observing that the pole of AEB with regard to the circle is at an infinite distance, draw through F, the pole of AB with regard to the conic, the line FL parallel to that diameter, QEq, of the circle which is perpendicular to AB.

The plane EFL will cut the conic in the diameter Pp, and the circle in the diameter Qq.

If pq, qP intersect FL in O and O', these two points will be vertices from which the conic can be projected into a circle, the centre of which is the projection of the point E.

Since $$FO : Fp :: Eq : Ep :: EA : Ep,$$

it follows that, for different positions of the plane through AB, FO is constant, so that O may be taken anywhere on the circle, centre F, in the plane through F perpendicular to the chord AEB.

Further, $$FO : FP :: EQ : EP :: EA : EP,$$

$$\therefore FO^2 : FP \,.\, Fp :: EA^2 : EP \,.\, Ep :: CD^2 : CP^2,$$

DCd being the semi-diameter of the conic which is conjugate to CP.

The length FO is thus determined when the position of the point E, inside the conic, is given, and, if we take as the vertex of projection any point O on the circle, centre F, as described above, the projection of the conic on any plane parallel to AEB and FO will be a circle.

If the conic is an ellipse, it follows that FO is equal to the ordinate FR, conjugate to Pp, of the hyperbola in the plane of the ellipse which has the same conjugate diameters PCp and DCd.

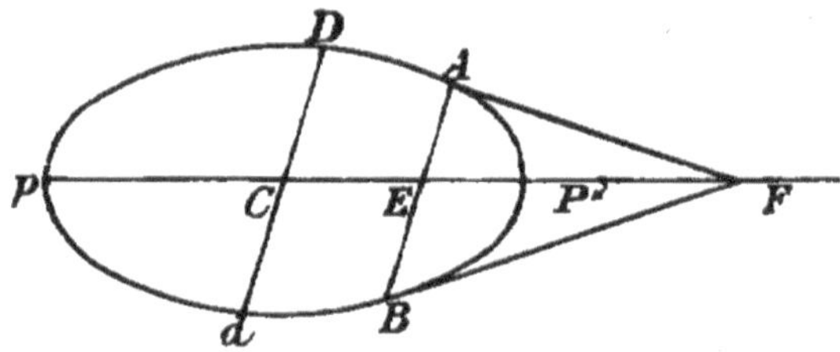

If the conic is an hyperbola, FO is equal to the ordinate FR of an ellipse in the plane of the hyperbola which has the same conjugate diameters PCp and DCd.

This hyperbola or this ellipse constructed outside the given conic may be called *the associated conic.*

If the conic is a parabola, the points O and O' are obtained by drawing lines through q and Q parallel to the axis of the parabola.

In this case,

$$FO^2 = Eq^2 = EA^2 = 4SP \,.\, PE = 4SP \,.\, PF,$$

so that the associated conic is a parabola.

If the conic is a circle, the associated conic is a rectangular hyperbola.

If the conic is an ellipse, the axes of which are indefinitely small, that is, if it is reduced to a point, the associated conic lapses into two straight lines, which are at right angles to each other if the point is the limit of a circle.

278. If the point E be outside the conic, or, in other words, if the polar of E intersect the conic, it is not possible to project the conic into a circle, so that the projection of E shall be the centre of the circle.

In this case the conic can be projected into a rectangular hyperbola, having the projection of the point E for its centre.

Let RU be the chord of contact of the tangents from E, and take any point O on the surface of the sphere of which RU is a diameter.

Then the projection of the conic from the vertex O on any plane parallel to ROU will be an hyperbola, and, since ROU is a right angle, it will be a rectangular hyperbola.

279. *If two conics in a plane are entirely exterior to each other, they can in general be projected, from the same vertex, into circles on the same plane.*

Draw four parallel tangents to the conics, and let F be the point of intersection of the diameters, PCp and QGq, joining the points of contact.

Also, let FR, FR' be the ordinates through F, parallel to the tangents, of the associated conics.

If F is so situated that these ordinates are equal, the locus of the vertices from which the two conics can be projected into circles will be the same, that is, it will be the circle of which F is the centre, and FR the length of the radius, in the plane through F perpendicular to FR.

In this case, taking any point O on the circle as the vertex, the two conics will be projected into circles on any plane parallel to the plane OFR, and the centres of the circles will be the projections of E and E', the respective poles of FR with regard to the conics.

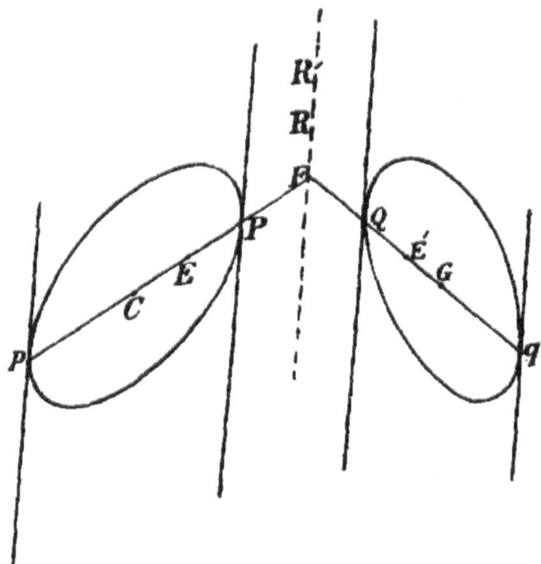

280. For different directions of the tangents, the points, F, R, R', will take up different positions, and for all directions of the tangents the loci of these points will be continuous curves.

The loci of R and R' will, in general, intersect each other; that is to say, there will be, in general, positions of F such that FR and FR' are equal.

Taking a particular case, let F be so situated that FR' is greater than FR; then taking F at the point where its locus meets the conic G, FR' vanishes, and therefore, between these two positions of F, there must be some position such that FR' is equal to FR.

We may observe that the locus of F passes through C and G, the centres of the two conics.

For, if CG is conjugate to the parallel tangents of the conic G, the point F is at C, and, if CG is conjugate to the parallel tangents of the conic C, the point F is at G.

When FR' is equal to FR, the line thus obtained is called by *Poncelet* the *Ideal Secant* of the two conics.

281. In a similar manner if one conic is entirely inside another they can, in general, be projected into circles, one of which will be inside the other.

Also two conics intersecting in two points may be projected into two intersecting circles.

Two conics intersecting in four points, or having contact at two points, cannot be projected into circles, but they can be projected into rectangular hyperbolas.

282. The method of projections enables us to extend to conics theorems which have been proved for a circle, and which involve, amongst other ideas, harmonic ranges, poles and polars, systems of collinear points, and systems of concurrent lines.

For instance, the theorems of Arts. 208 and 210 are easily proved for a circle, and by this method are at once extended to conics.

Take as another instance Pascal's theorem, that *the opposite sides of any hexagon inscribed in a conic intersect in three collinear points.*

If this be proved for a circle, the method of conical projection at once shews that it is true for any conic.

The following very elementary proof of the theorem for a circle is given in *Catalan's Théorèmes et Problèmes de Géométrie Elémentaire.*

Let $ABCDEF$ be the hexagon, and let AB and ED meet in G, BC and FE in H, FA and DC in K.

Also let ED meet BC in M and AF in N, and let BC meet AF in L.

Then we have the relations,

$$LA \,.\, LF = LB \,.\, LC, MC \,.\, MB = MD \,.\, ME,$$

$$NE \,.\, ND = NF \,.\, NA.$$

Also, the triangle LMN being cut by the three transversals AG, DK, FH, we have the relations,

$$LB \,.\, MG \,.\, NA = LA \,.\, MB \,.\, NG$$

$$LC \,.\, MD \,.\, NK = LK \,.\, MC \,.\, ND$$

$$LH \,.\, ME \,.\, NF = LF \,.\, MH \,.\, NE.$$

Multiplying together these six equalities, taking account of the relations previously stated, and cutting out the factors common to the two products, we obtain

$$LH \,.\, MG \,.\, NK = LK \,.\, MH \,.\, NG;$$

$\therefore$ G, H, K are collinear.

Brianchon's theorem that, *if a hexagon circumscribe a conic, the three opposite diagonals are concurrent* is proved at once by observing that it is the reciprocal polar of Pascal's theorem.

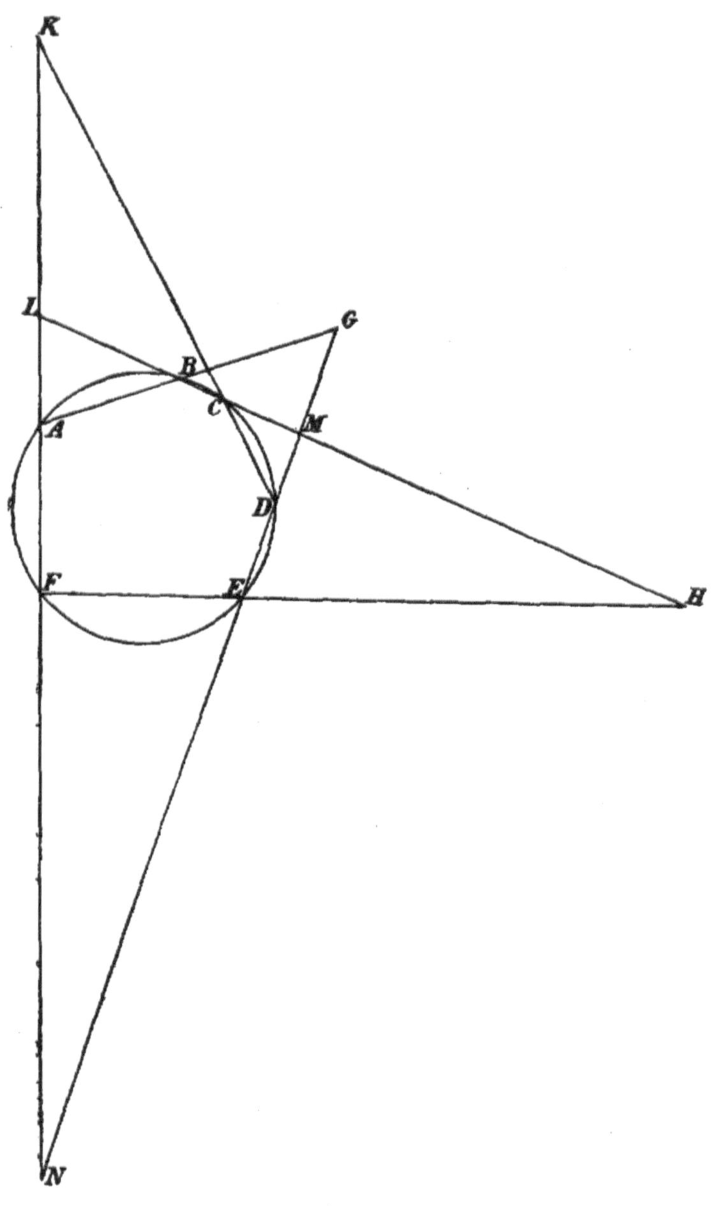
K
L
G
B
C
A
M
D
F
E
H
N

283. *Stereographic and Gnomonic Projections.*

If a point on the surface of a sphere be taken as the vertex of projection, and if the plane of projection be parallel to the tangent plane at the point, the projection of any figure drawn on the surface of the sphere is called its stereographic projection.

If however the centre of the sphere be taken as the vertex of projection, and any plane be taken as the plane of projection, the projection of any figure drawn on the surface of the sphere is called its gnomonic projection.

The stereographic projection of a circle drawn on the surface of the sphere is a circle; for it can be easily shewn that it is a subcontrary section of the oblique cone formed by the vertex of projection and the circle on the sphere.

The gnomonic projection of a circle on the sphere is obviously a conic.

These projections are sometimes described in treatises on Astronomy, and in these treatises the vertex for stereographic projection is taken at the south pole of the earth, and, for gnomonic projection, at the centre of the earth; and, in both cases, the plane of projection is taken parallel to the plane of the equator.

284. It will be seen that the discussions which are given in this chapter are confined entirely to cases of *real* projection.

The chapter is intended to be simply an introduction to a large and important subject.

The method of conical projections is due to Poncelet, and is worked out with great fulness and elaboration in his work entitled, *Traité des Propriétés Projectives des Figures* (Second edition, 1865, in two quarto volumes).

In this work Poncelet extends the domain of pure geometry by the interpretation and use of the law of continuity, and, as one of its applications, by the introduction of the imaginary chord of intersection, or, as it is called by Poncelet, the ideal secant of two conics.

Amongst English writers, the student will find valuable chapters on projections in *Salmon's Conics*, and in the large work on the *Geometry of Conics, by Dr C. Taylor, the Master of St John's College, Cambridge.*

There is also an important work by *Cremona, on Projective Geometry*, which has been translated by *Leudesdorf* (Second edition, 1893).

MISCELLANEOUS PROBLEMS. II.

1. If two conics have the same directrix, their common points are concyclic.

2. If a focal chord of a parabola is bisected in V and the line perpendicular to it through V meets the axis in G, SG is half the chord.

3. If the perpendicular to CP from a point P of an ellipse meets the auxiliary circle in Q, PQ varies as PN.

4. AA' and BB' are the axes, and S is one of the foci of an ellipse; if a parabola is described with S as focus and passing through B and B', its vertex bisects SA or SA'.

5. Tangents to an ellipse at P, p intersect on an axis; if the perpendicular from p on the tangent at P intersects CP in L, the locus of L is a similar ellipse.

6. The normal to a hyperbola at P meets the axes in G and g respectively. Prove that the circle circumscribing SPG is touched Sg.

7. If a tangent to an ellipse meets a pair of conjugate diameters in points equidistant from the centre, the locus of the points is a circle.

8. If ellipses are described on AB as diameter, touching BC, the points of contact of tangents from C are on a straight line.

9. If Pl, Pm be drawn perpendicular to CL, CM respectively, shew that the centre of the circle Plm lies on a fixed hyperbola.

10. PSQ, PHR are focal chords of an ellipse, QT, RT the tangents at Q and R. Shew that PT is the normal at P.

11. A, B are two fixed points. Through them a system of circles is drawn. Through A draw any two lines meeting the circles in the points C_1D_1, C_2D_2, &c. Shew that the lines CD all touch a parabola, focus B, which also touches the lines AC, AD.

12. From any two points A, B on an ellipse four lines are drawn to the foci S, H. Shew that $SA \,.\, HB$ and $SB \,.\, HA$ are to one another as the squares of the perpendiculars from a focus on the tangents at A and B.

13. If two points of a conic and the angle subtended by these points at the focus are given, the line joining the focus with the intersection of the tangents always passes through a fixed point.

14. If normals to an ellipse are drawn at the extremities of chords parallel to one of the equi-conjugate diameters, pairs of such normals intersect on the line through the centre perpendicular to the other diameter.

15. From the point in which the tangent at any point P of a hyperbola cuts either asymptote perpendiculars are dropped upon the axes. Prove that the line joining the feet of these perpendiculars passes through P.

16. Tangents are drawn to an ellipse parallel to conjugate diameters of a second given ellipse. Shew that the locus of their intersection is an ellipse similar and similarly situated to the second ellipse.

17. A focus of a conic inscribed in a triangle being given, find the points of contact.

18. The normals at P and Q, the ends of a focal chord PSQ, intersect in K, and KN is perpendicular to PQ; prove that NP and SQ are equal.

19. If CR, SY, HZ be perpendiculars upon the tangent at a point P such that $CR = CS$, prove that R lies on the tangent at B, and that the perpendicular from R on SH will divide it into two parts equal to SY, HZ respectively.

20. If a parabola, having its focus coincident with one of the foci of an ellipse, touches the conjugate axis of the ellipse, a common tangent to the ellipse and parabola will subtend a right angle at the focus.

21. Two tangents TP and TQ are drawn to an ellipse, and any chord TRS is drawn, V being the middle point of the intercepted part; QV meets the ellipse in P'; prove that PP' is parallel to ST.

22. If S, S' are the foci of an ellipse and SY, $S'Y'$ the perpendiculars on any tangent, XY, $X'Y'$ meet on the minor axis, and, if PN is the ordinate of P, NY and NY' are perpendicular to XY and $X'Y'$ respectively.

23. A circle through the centre of a rectangular hyperbola cuts the curve in the points A, B, C, D. Prove that the circle circumscribing the triangle formed by the tangents at A, B, C passes through the centre of the hyperbola.

24. If the tangent at a point P of an ellipse meets any pair of parallel tangents in M, N, and if the circle on MN as diameter meets the normal at P in K, L,

then KL is equal to DCD', and CK, CL are equal to the sum and difference of the semi-axes.

25. From a point O two tangents OA, OB are drawn to a parabola meeting any diameter in P, Q. Prove that the lines OP, OQ are similarly divided by the points of contact, but one internally, the other externally.

26. If S, H be the foci of an ellipse, and SP, HQ be parallel radii vectores drawn towards the same parts, prove that the tangents to the ellipse at P, Q intersect on a fixed circle.

27. If an ellipse be inscribed in a quadrilateral so that one focus S is equidistant from the four vertices, the other focus must be at the intersection H of the diagonals.

28. P is a point on a circle whose centre is Q; through P a series of rectangular hyperbolas are described having Q for their centre of curvature at P. Prove that the locus of their centres is a circle with diameter of length PQ.

29. Two cones which have a common vertex, their axes at right angles, and their vertical angles supplementary, are intersected by a plane at right angles to the plane of their axes. Prove that the distances of either focus of the elliptic section from the foci of the hyperbolic section are equal respectively to the distance from the vertex of the ends of the transverse axis of each, and that the sum of the squares on the semi-conjugate axes is equal to the rectangle contained by those distances.

30. Two plane sections of a cone which are not parallel are such that a focus of each and the vertex of the cone lie on a straight line. Shew that the angle included by any pair of focal chords of one section is equal to that contained by the corresponding focal chords of the other section, corresponding chords being the projections of each other with respect to the vertex.

31. If PP', QQ' be chords normal to a conic at P and Q, and also at right angles to each other, then will PQ be parallel to $P'Q'$.

32. A system of conics have a common focus S and a common directrix corresponding to S. A fixed straight line through S intersects the conics, and at the points of intersection normals are drawn. Prove that these normals are all tangents to a parabola.

33. If two confocal conics intersect, prove that the centre of curvature of either curve at a point of intersection is the pole of the tangent at that point with regard to the other curve.

34. A chord of a conic whose pole is O meets the directrices in R and R'; if SR and HR' meet in O', prove that the minor axis bisects OO'.

35. TQ and TR, tangents to a parabola, meet the tangent at P in X and Y, and TU is drawn parallel to the axis, meeting the parabola in U. Prove that the tangent at U passes through the middle point of XY, and that, if S is the focus,

$$XY^2 = 4SP \,.\, TU.$$

36. The foot of the directrix which corresponds to S is X, and XY meets the minor axis in T; CV is the perpendicular from the centre on the tangent at P. Prove that, if $CP = CS$, then $CV = VT$.

37. A is a given point in the plane of a given circle, and ABC a given angle. If B moves round the circumference of the circle, prove that, for different values of the angle ABC, the envelopes of BC are similar conics, and that all their directrices pass through one or other of two fixed points.

38. If AA' is the transverse axis of an ellipse, and if Y, Y' are the feet of the perpendiculars let fall from the foci on the tangent at any point of the curve, prove that the locus of the point of intersection of AY and $A'Y'$ is an ellipse.

39. The tangent at a point P of an hyperbola cuts the asymptotes in L and L', and another hyperbola having the same asymptotes bisects PL and PL'. Prove that it intersects CP in a point p such that

$$Cp^2 : CP^2 :: 3 : 4.$$

The chord QR, joining a point R on an asymptote with a point Q on the corresponding branch of the first hyperbola, intersects the second hyperbola in E; if QR move off parallel to itself to infinity, prove that, ultimately $RE : EQ :: 3 : 1$.

40. Tangents are drawn to a rectangular hyperbola from a point T in the transverse axis, meeting the tangents at the vertices in Q and Q'. Prove that QQ' touches the auxiliary circle at a point R such that RT bisects the angle QTQ'.

41. Tangents from a point T touch the curve at P and Q; if PQ meet the directrices in R and R', PR and QR' subtend equal angles at T.

42. The straight lines joining any point to the intersections of its polar with the directrices touch a conic confocal with the given one.

43. If a point moves in a plane so that the sum or difference of its distances from two fixed points, one in the given plane and the other external to it, is constant, it will describe a conic, the section of a right cone whose vertex is the given external point.

44. In the construction of Art. 241 prove that CK' and CK are respectively equal to the sum and difference of the semi-axes.

45. Given a tangent to an ellipse, its point of contact, and the director circle, construct the ellipse.

46. If the tangent at any point P of an ellipse meet the auxiliary circle in Q', R', and if Q, R be the corresponding points on the ellipse, the tangents at Q and R pass through the point P' on the auxiliary circle corresponding to P.

47. In the ellipse $PDP'D'$, $P'HCSPX$ and DCD' are conjugate diameters; CH is equal to CS, and the polar of S passes through a point X on $P'P$ produced. If DX is drawn cutting the ellipse in Q, prove that HD is parallel to SQ.

48. If T is the pole of a chord of a conic, and F the intersection of the chord with the directrix, TSF is a right angle.

49. The polar of the middle point of a normal chord of a parabola meets the focal vector to the point of intersection of the chord with the directrix on the normal at the further end of the chord.

50. OP, OQ touch a parabola at P, Q; the tangent at R meets OP, OQ in S, T; if V is the intersection of PT, SQ, O, R, V are collinear.

51. If from any point A a straight line AEK be drawn parallel to an asymptote of an hyperbola, and meeting the polar of A in K and the curve in E, shew that $AE = EK$.

52. If a chord PQ of a parabola, whose pole is T, cut the directrix in F, the tangents from F bisect the angle PFT and its supplement.

53. A parabola, focus S, touches the three sides of a triangle ABC, bisecting the base BC in D; prove that AS is a fourth proportional to AD, AB, and AC.

54. A focal chord PSQ is drawn to a conic of which C is the centre; the tangents and normals at P and Q intersect in T and K respectively; shew that ST, SP, SK, SC form an harmonic pencil.

55. PCP' is any diameter of an ellipse. The tangents at any two points D and E intersect in F. PE, $P'D$ intersect in G. Shew that FG is parallel to the diameter conjugate to PCP'.

56. A conic section is circumscribed by a quadrilateral $ABCD$: A is joined to the points of contact of CB, CD; and C to the points of contact of AB, AD; prove that BD is a diagonal of the interior quadrilateral thus formed.

57. A parabola touches the three lines CB, CA, AB in P, Q, R, and through R a line parallel to the axis meets RQ in E; shew that $ABEC$ is a parallelogram.

58. If a series of conics be inscribed in a given quadrilateral, shew that their centres lie on a fixed straight line.

Shew also that this line passes through the middle points of the diagonals.

59. Four points A, B, C, D are taken, no three of which lie in a straight line, and joined in every possible way; and with another point as focus four conics are described touching respectively the sides of the triangles BCD, CDA, DAB, ABC; prove that the four conics have a common tangent.

60. If the diagonals of a quadrilateral circumscribing a conic intersect in a focus, they are at right angles to one another, and the third diagonal is the corresponding directrix.

61. An ellipse and parabola have the same focus and directrix; tangents are drawn to the ellipse at the extremities of the major axis; shew that the diagonals of the quadrilateral formed by the four points where these tangents cut the parabola intersect in the common focus, and pass through the extremities of the minor axis of the ellipse.

62. Three chords of a circle pass through a point on the circumference; with this point as focus and the chords as axes three parabolas are described whose parameters are inversely proportional to the chords; prove that the common tangents to the parabolas, taken two and two, meet in a point.

63. A circle is described touching the asymptotes of an hyperbola and having its centre at the focus. A tangent to this circle cuts the directrix in F, and has its pole with regard to the hyperbola at T. Prove that TF touches the circle.

64. Two conics have a common focus: their corresponding directrices will intersect on their common chord, at a point whose focal distance is at right angles to that of the intersection of their common tangents. Also the parts into which either

common tangent is divided by their common chord will subtend equal angles at the common focus.

If the conics are parabolas, the inclination of their axes will be the angle subtended by the common tangent at the common focus.

65. The tangent at the point P of an hyperbola meets the directrix in Q; another point R is taken on the directrix such that QR subtends at the focus an angle equal to that between the transverse axis and an asymptote; prove that the envelope of RP is a parabola.

66. If an hyperbola passes through the angular points of an equilateral triangle and has the centre of the circumscribing circle as focus, its eccentricity is the ratio of 4 to 3, and its latus rectum is one-third of the diameter of the circle.

67. An isosceles triangle is circumscribed to a parabola; prove that the three sides and the three chords of contact intersect the directrix in five points, such that the distance between any two successive points subtends the same angle at the focus.

68. Tangents are drawn at two points P, P' on an ellipse. If any tangent be drawn meeting those at P, P' in R, R', shew that the line bisecting the angle RSR' intersects RR' on a fixed tangent to the ellipse.

69. The chords of a conic which subtend the same angle at the focus all touch another conic having the same focus and directrix.

70. Two conics have a common focus S and a common directrix, and tangents TP, TP' are drawn to one from any point on the other and meet the directrix in F and F'. Prove that the angles PSF', $P'SF$ are equal and constant.

71. A rectangular hyperbola circumscribes a triangle ABC; if D, E, F are the feet of the perpendiculars from A, B, C on the opposite sides, the loci of the poles of the sides of the triangle ABC are the lines EF, FD, DE.

72. If two of the sides of a triangle, inscribed in a conic, pass through fixed points, the envelope of the third side is a conic.

73. If two circles be inscribed in a conic, and tangents be drawn to the circles from any point in the conic, the sum or difference of these tangents is constant, according as the point does or does not lie between the two chords of contact.

74. The four common tangents of two conics intersect two and two on the sides of the common self-conjugate triangle of the conics.

75. Prove that a right cylinder, upon a given elliptic base, can be cut in two ways so that the curve of section may be a circle; and that a sphere can always be drawn through any two circular sections of opposite systems.

76. An ellipse revolves about its major axis, and planes are drawn through a focus cutting the surface thus formed. Prove that the locus of the centres of the different sections is a surface formed by the revolution of an ellipse about CS where C or S are respectively the centre and focus of the original ellipse.

77. Given five tangents to a conic, find, by aid of Brianchon's theorem, the points of contact.

78. The alternate angular points of any pentagon $ABCDE$ are joined, thus forming another pentagon whose corresponding angular points are a, b, c, d, e; Aa, Bb, Cc, Dd, Ee are joined and produced to meet the opposite sides of $ABCDE$ in α, β, γ, ϵ; shew that if A be joined with the middle point of $\gamma\delta$, B with the middle point of $\delta\epsilon$, &c., these five lines meet in a point.

79. If a conic be inscribed in a triangle, the lines joining the angular points to the points of contact of the opposite sides are concurrent.

80. If a quadrilateral circumscribe a conic, the intersection of the lines joining opposite points of contact is the same as the intersection of the diagonals.

81. ABC is a triangle, and D, E, F the middle points of the sides. Shew that any two similar and similarly situated ellipses one circumscribing DEF and the other inscribed in ABC will touch each other.

82. AB is a chord of a conic. The tangents at A and B meet in T. Through B a straight line is drawn meeting the conic in C and AT in P. The tangent to the conic at C meets AT in Q. Prove that $TPQA$ is a harmonic range.

83. Pp, Qq, Rr, Ss are four concurrent chords of a conic; shew that a conic can be drawn touching SR, RQ, QP, sr, rq, qp.

84. If two sections of a right cone have a common directrix, the latera recta are in the ratio of the eccentricities.

85. $ABCD$ is a parallelogram and a conic is described to touch its four sides. If S is a focus of this conic and if with S as focus a parabola is described to touch AB and BC, the axis of the parabola passes through D.

86. If from a point O tangents be drawn to two conics S and S', and if the tangents to S be conjugate with respect to S', prove that the tangents to S' are conjugate with respect to S.

87. If a triangle is self-conjugate with respect to each of a series of parabolas, the lines joining the middle points of its sides will be tangents; all the directrices will pass through O, the centre of the circumscribing circle; and the focal chords, which are the polars of O, will all touch an ellipse inscribed in the given triangle which has the nine-point circle for its auxiliary circle.

88. If a triangle can be drawn so as to be inscribed in one given conic and circumscribed about another given conic, an infinite number of such triangles can be drawn.

89. Prove that the stereographic projection of a series of parallel circles on a sphere is a series of coaxal circles, the limiting points of which are the projections of the poles of the circles.

90. Through the six points of intersection of a conic with the sides of a triangle straight lines are drawn to the opposite angular points; if three of these lines are concurrent the other three are also concurrent.

91. Prove that the asymptotes of an hyperbola, and a pair of conjugate diameters form an harmonic range, and that the system of pairs of conjugate diameters is a pencil in involution.

92. If two concentric conics have the directions of two pairs of conjugate diameters the same, then the directions are the same for every pair.

93. If two concentric conics have all pairs of conjugate diameters in the same directions, and have a common point, they coincide entirely.

94. If two conics have two common self-conjugate triangles with the same vertex, which is interior to both, they cannot intersect in any point without entirely coinciding.

95. If two conics in space whose planes intersect in a line which does not cut either conic, and if on this line there are four points, P, P', Q, Q', such that the polars of P with regard to the conics both pass through P', and that the polars of Q both pass through Q', then either conic can be projected into the other in two ways.

www.ingramcontent.com/pod-product-compliance
Lightning Source LLC
LaVergne TN
LVHW101914190826
846094LV00001B/4

* 9 7 8 9 3 9 3 9 7 1 5 2 4 *